Greening Brazil

Kathryn Hochstetler & Margaret E. Keck

Greening Brazil

Environmental Activism in State and Society

DUKE UNIVERSITY PRESS
Durham & London • 2007

© 2007 Duke University Press
All rights reserved
Printed in the United States of America on acid-free paper
Designed by Jennifer Hill
Typeset in Dante by Keystone Typesetting, Inc.

Library of Congress Cataloging-in-Publication Data appear
on the last printed page of this book.

We dedicate this book to the next generation—
Melissa and Laura—

and the next wave of
Brazilian environmentalism

Contents

LIST OF TABLES	viii
PREFACE	ix
LIST OF ACRONYMS AND ORGANIZATIONS	xv
Introduction	1
1	
Building Environmental Institutions	
National Environmental Politics and Policy	23
2	
National Environmental Activism	
The Changing Terms of Engagement	63
3	
From Protest to Project	
The Third Wave of Environmental Activism	97
4	
Amazônia	140
5	
From Pollution Control to Sustainable Cities	186
Conclusion	223
APPENDIX: LIST OF INTERVIEWS	231
NOTES	239
BIBLIOGRAPHY	249
INDEX	273

Tables

1.1
FEDERAL ENVIRONMENTAL INSTITUTIONS IN BRAZIL 38

1.2
FEDERAL ENVIRONMENTAL SPENDING, 1993–2000, 2003–2005 41

3.1
EMPLOYMENT IN NGOS FOR THE PROTECTION OF THE ENVIRONMENT AND OF ANIMALS, 2002 107

3.2
ASSOCIATIONS FOR THE PROTECTION OF THE ENVIRONMENT AND ANIMALS, 2002 131

4.1
FEDERAL CONSERVATION UNITS IN AMAZÔNIA 170

Preface

This book has been a long time in the making. Our object of study—origins, strategies, and the political "fit" of environmentalism and environmentalists in Brazil—was more elusive than either of us expected. Neither a single social movement, nor a policy area, nor even a clearly bounded corpus of ideas, environmentalism in a large developing country was a moving target, as it had to be to make sense in its home territory; nonetheless, by refusing to fit the theoretical pigeonholes into which we wanted to place it, it made our task harder. This research started out in 1989 as two separate projects, both of which long ago produced book manuscripts with which neither of us was fully satisfied. We became involved in other projects. The idea for this book resulted from a serendipitous meeting in June 2000 at the airport in São Paulo, where it occurred to us to put our books together. Several years and many conversations later, there is almost nothing left from the original manuscripts, but we hope the combined result is closer to what each of us wishes she had written in the first place.

Obviously this book does not reflect a classic research design with a well-formulated initial hypothesis, investigated in the field and analyzed and reported promptly. Instead it is the synthetic product of a whole series of research efforts, carried out independently by different scholars with somewhat different preoccupations over a number of years. In rethinking, resituating, and rewriting the manuscript, we tried to give it conceptual and narrative unity, but there will inevitably be signs of its several origins. Each of us has built continuously on the fieldwork that she began fifteen years ago, but we now see that fieldwork in the light of continuing research on environmental politics and research in Brazil. We have seen the birth and death of organizations and institutions, witnessed life-cycle and other changes in the Brazilian activists we have known over the years, and been present during key events and processes of the political and economic changes that Brazil has undergone since the military left power in the mid-1980s. In the meantime, we have both developed close friendships and rich collegial relations with Brazilians, and have learned enough to doubt certainties.

From this long-gestated research, we drew three major lessons that shape this book. First, we needed to pay more attention to relations between domestic and international actors, conscious that most international portrayals miss most of the story, and often get the dynamics wrong on the parts they do capture. That each of us collaborated on books about transnational relations in the meantime is obviously relevant as well. Second, a longer timeline highlights just how thoroughly embedded environmental politics is in a larger set of political, social, and economic relations, domestically as well as internationally. Without a broader understanding of Brazilian politics more generally—the impact of democratization, federalism, and the high levels of informality that challenge the implementation and institutionalization of policies—it is impossible to understand environmental politics. Finally, our early images of environmental activists in civil society pressuring state institutions for changes in policies and behavior have given way to a recognition that activists labor mightily inside the state as well as outside it, and that an accurate portrayal requires keeping both sites of struggles in view. The Introduction relates these lessons to some of the existing debates on international and comparative environmental politics, and on the relationships between the two.

This is a largely descriptive work—informed by and in constant dialogue with theory, but not intended as a test of any one in particular, in the tradition of grounded theory. The book's theoretical ambitions are modest, aiming to (1) produce a more nuanced view of the kinds of interactions that shape a multilevel governance of the environment in Brazil; (2) demonstrate the importance of studying particular policy areas like the environment within a broader political context that recognizes interactions between different levels of political institutions and among state and society actors, each of which has multiple commitments and connections; and (3) identify some of the patterns by which committed actors inside and outside the state attempt to make, maintain, or block policies against powerful but dispersed opponents, through blocking or enabling networks.

In the Introduction we promise to tell the "inside" story of Brazilian environmental politics that the transnationalized version of the story leaves out. This is an audacious claim for a pair of foreigners, especially in a context where both the quality and quantity of national scholarship on environmental politics are high. The environmental sections of the National Association of Graduate Study and Research in the Social Sciences (ANPOCS), the contributors to journals like *Ambiente e Sociedade* (Environment and Society), the social science participants in the Brazilian Society for the Advancement of Science (SBPC), and others have already provided important insights into Brazilian environmental politics, and we cite their work repeatedly in this book.

At the same time, we believe there are additional insights that we can bring as scholars who do not live in Brazil but return there repeatedly—in Margaret Keck's case for the last twenty-five years and in Kathryn Hochstetler's for fifteen. Leaving Brazil to experience environmental debates as citizens in the United States, do research in other South American countries, and observe transnational interactions gives us a comparative vantage point that clarifies Brazil's unique qualities and commonalities with other cases. Absences also make some of the transition points more noticeable: when we were here a year ago, that organization still existed, this option still seemed possible while now it does not, these allies were enemies (or vice versa), and so on. Overall this is not a better vantage point, but it is a different one.

Returning regularly to Brazil also gives us a vantage point not shared by those whose view of Brazilian environmental politics comes from international settings and the international media. For reasons that we discuss directly in the Introduction and indirectly throughout the book, the positions taken by Brazilians in international settings are often themselves not fully reflective of domestic environmental developments. The *longue durée*, the regular monthly meetings, and much more are simply not visible even to many Brazilians. They are critical for the unfolding of Brazilian environmental politics, but will rarely make headlines.

We have studied the headlines and the monthly meetings, using a range of methods that combined hundreds of semi-structured and open-ended interviews, extensive participant observation, documentary and archival research, and analysis of some quantitative data. They have also involved revisits to many of our research sites, some repeatedly. Just for accompanying the Brazilian preparations for the Earth Summit, for example, Hochstetler attended three national meetings of the Brazilian NGO Forum (including one of the first where foreign NGOs were present in a related meeting), three state-level preparatory meetings, and at least a dozen additional meetings of São Paulo's statewide association of environmental groups, APEDEMA. She also accompanied the month-long government negotiations in New York of the Fourth Preparatory Conference, a South American gathering sponsored by Friends of the Earth, and the national preparatory process in Venezuela (providing them with documents and information from Brazil, not otherwise available). Keck attended two of the Brazilian national meetings, and we in fact met for the first time at the third meeting of the Brazilian NGO Forum in October 1990. Along the way, we have traveled to sixteen (Hochstetler) and nineteen (Keck) of Brazil's twenty-seven subnational units. The long time horizon and broad geographic grounding has advantages and disadvantages. Along with the advantage of deeper and more nuanced appreciation of the processes at work goes the frustration when, at the end of the period, we know so many questions we should have asked at the beginning. Our research trajectories have been similar.

In 1990 Margaret Keck set out to investigate the developing linkages between environmentalists and social movements struggling for material improvements in their living conditions. She resisted the categories of "new" and "old" social movements prevalent in the North, finding instead different combinations of demands for a better life and for a different life in a wide range of movements. Fascinated by the hybrid discourse of "social environmentalism" emerging from groups like the Acre rubber tappers, she set out to look for its urban equivalent. But although many urban environmentalists used the term "social environmentalism," their practice—and their histories—told different stories. Her detailed study of environmental struggles over São Paulo's water quality, involving developments around the Billings and Guarapiranga dams, led her to focus on networks of activists in state and society as they enabled or obstructed particular state policies (Keck 2001). Her study of the transnational relations surrounding efforts to enforce the environmental provisions of the World Bank's Planafloro loan in Rondônia (Keck 1998) stimulated her collaboration with Kathryn Sikkink on transnational advocacy networks (Keck and Sikkink 1998a).

Kathryn Hochstetler also began with a focus on environmental movements, comparing the ways that such movements in Brazil and Venezuela balanced strategic and identity-based considerations as they sought to bring environmental concerns into political systems preoccupied with both development and democratization. Her later work continued to place Brazilian environmental actors—in state and society—in a comparative and international context, looking at a series of mobilizations around the La Plata River basin and the Mercosul (Mercosur) free trade area as well as in United Nations conferences. This international and comparative work oriented her toward seeing Brazil as an environmental innovator in its regional context, even as its ongoing environmental gaps and failures were also evident. Both sides, to her, justify a closer, deeper look at this environmental puzzle. This book is that.

A project like this one inevitably garners more debts than can be listed, much less repaid. Our largest debt is to the many Brazilians who have given generously of their time and papers to explain Brazilian environmental politics to us, in formal interviews and by simply letting us observe their activities. Above all, we appreciate how interesting and articulate they have been, and how hard they have worked to find creative solutions to often daunting problems.

We also owe more specific thanks. Maria Helena Antuniassi at CERU of the University of São Paulo shared with Kathryn Hochstetler an important set of early documents and interviews from the environmental movement. Hochstetler held research affiliations at CEBRAP in São Paulo, IUPERJ in Rio de Janeiro, and the University of Brasília during different parts of this research. An important portion of her writing time was hosted by the Centre for Brazilian Studies of Oxford University, which sponsored a conference, "Forests, Cities, Climate Change and Poverty: New Perspectives on Environmental Politics in Brazil," that allowed us to get helpful comments on an early set of chapter drafts. Hochstetler's research was funded by grants from the Institute for the Study of World Politics, the Midwest Universities Consortium for International Affairs, a Fulbright faculty research grant, and the Career Enhancement and College of Liberal Arts Professional Development funds of Colorado State University. While Hochstetler is grateful for all of this support, she would especially like to thank her colleagues and students at Colorado State University, who were interested supporters of this project for its entire duration.

Margaret Keck thanks the Yale Center for International and Area Studies for support in the very early stages of this project, as well as the Yale and Johns Hopkins students who have participated in her seminar on environment and development over the last fifteen years, the Yale Agrarian Studies Seminar, and her colleagues in the Yale and Johns Hopkins political science departments. During portions of the research she was affiliated with CEDEC in São Paulo and was generously welcomed by IPHAE in Porto Velho. She was lucky to have as research assistants at different moments Biorn Maybury-Lewis, Denise Campelo, and Cristina Saliba. Beyond assistance with the research itself, Cristina Saliba provided a home away from home, steadfast friendship, and boundless generosity during much of the time this research was being done. Keck's research was funded by grants from the Howard Heinz Endowment / Center for Latin American Studies, University of Pittsburgh; the Joint Committee on Latin American Studies and the Advanced Fellowship in Foreign Policy Studies of the Social Science Research Council and the American Council of Learned Societies, with funds provided by the Ford Foundation; the John D. and Catherine T. MacArthur Foundation; and research funds provided by Johns Hopkins University.

Innumerable people have commented on our research projects or discussed ideas with us in ways that have been partially incorporated into this book. We resolve the impossibility of naming all of them by naming none, but appreciating each one more than we can say. In addition, we thank the following people for helpful comments on the chapters and arguments of the book itself (Brazilian style, they are alphabetized by first name): Alberto Lourenço, Andy Hurrell, Charles Wood, Jonathan Fox, José Augusto Pádua, Lesley McAllister, Lupe Rodrigues, Mary Allegretti, Rebecca Abers, Sylvia Tesh, Timmons Roberts, and several anonymous reviewers. We are, of course, responsible for all remaining errors of fact and interpretation. At Duke, Valerie Millholland has been a prompt and encouraging editor.

In a project of this length, it is remarkable that our partners—Roger Hoover and Larry Wright—have been present and supportive through the entire process. Melissa Wright, born not long after the research began, and Laura Wright, born two years later, tolerated Keck's absences and grew up along with the book. We thank all for their patience. Now, other things can get done.

Acronyms and Organizations

ABEMA Associação Brasileira de Entidades de Meio Ambiente [Brazilian Association of Environmental Agencies]

ABES Associação Brasileira de Engenharia Sanitária [Brazilian Association of Sanitary Engineers]

ABONG Associação Brasileira de ONGs [Brazilian Association of NGOs]

ABRAE Associação Brasileira dos Expostos ao Amiento [Brazilian Association of Those Exposed to Asbestos]

AEA Associação dos Engenheiros Agrônomos [Association of Agronomists]

AEBA Association of Ex-Scholarship Holders in Germany

ACPO Associação de Combate aos POPs [Association against Persistent Organic Pollutants]

AGAPAN Associação Gaúcha de Proteção ao Ambiente Natural [Gaucho Association for Protection of the Natural Environment]

APEDEMA Assembléia Permanente de Entidades em Defesa do Meio Ambiente [Permanent Assembly of Environmental Defense Organizations]

APPN Associação Paulista de Proteção da Natureza [São Paulo Association for Nature Protection]

ARENA Aliança Nacional Renovadora [National Renewal Alliance (Party)]

BNDES Banco Nacional de Desenvolvimento [National Economic Development Bank]

Brazilian NGO Forum Fórum de ONGs Brasileiras Preparatório para a Conferência da Sociedade Civil sobre Meio Ambiente e Desenvolvimento [Brazilian NGO Forum in Preparation for the Civil Society Conference on Environment and Development (UNCED)]

CDPC Comissão de Defesa do Patrimônio da Comunidade [Commission to Defend the Community Heritage]

CEDI Centro Ecumênico de Documentação e Informação

CETESB Companhia de Tecnologia de Saneamento Ambiental [Environmental Sanitation Technology Company (São Paulo)]

CGT Central Geral dos Trabalhadores [General Workers' Confederation]

CIEC Coordenação Interestadual de Ecologistas para a Constituinte [Interstate Ecological Coordination for the Constituent Assembly]

CNDDA Campanha Nacional para a Defesa e pelo Desenvolvimento da Amazônia [National Campaign for the Defense and Development of the Amazon]

CNI Confederação Nacional da Industria

CNS Conselho Nacional dos Seringueiros [National Rubber Tappers Council]

CONAMA Conselho Nacional do Meio Ambiente [National Environmental Council]

CONSEMA Conselho Estadual do Meio Ambiente [State Environment Council (São Paulo)]

CUT Central Única dos Trabalhadores [Unified Workers' Confederation]

CVRD Companhia Vale do Rio Doce [Vale do Rio Doce Company]

Eco '92 United Nations Conference on Environment and Development

ENEAA Encontro Nacional de Entidades Autónomas Ambientalistas [National Encounter of Autonomous Environmental Organizations]

FASE Federação de Órgãos para a Assistência Social e Educacional [Federation of Organs for Social and Educational Assistance]

FBCN Fundação Brasileira de Conservação de Natureza [Brazilian Foundation for the Conservation of Nature]

FBOMS Fórum Brasileiro de ONGs e Movimentos Sociais para o Meio Ambiente e o Desenvolvimento [Forum of Brazilian NGOs and Social Movements for Environment and Development]

FEEMA Fundação Estadual de Engenharia do Meio Ambiente [State Foundation for Environmental Engineering (Rio de Janeiro)]

FINEP Financiadora de Estudos e Projetos [Funding Agency for Studies and Projects]

FORUM Fórum de ONGs Brasileiras Preparatório para a Conferência da Sociedade Civil sobre Meio Ambiente e Desenvolvimento [Brazilian NGO Forum in Preparation for the Civil Society Conference on Environment and Development]

FVA Fundacão Vitória Amazônica

GM genetically modified

G-7 Group of Seven (wealthiest countries)

IBAMA Instituto Brasileiro do Meio Ambiente e dos Recursos Naturais Renováveis [Brazilian Institute of the Environment and Natural Resources]

IBASE Instituto Brasileiro de Análises Sociais e Econômicas [Brazilian Institute of Social and Economic Analyses]

IBDF Instituto Brasileiro de Desenvolvimento Florestal [Brazilian Institute for Forestry Development]

IBGE Instituto Brasileiro de Geografia e Estatística [Brazilian Geographical and Statistical Institute]
IDB Interamerican Development Bank
IEA Instituto de Estudos Amazônicos
IFC International Facilitating Committee
INCRA Instituto Nacional de Colonização e Reforma Agrária [Colonization and Land Reform Institute]
INPA Instituto de Pesquisas da Amazônia [Institute for Amazonian Research]
IPEN Instituto de Pesquisas Energêticas e Nucleares [Institute for Energy and Nuclear Research]
IUCN International Union for the Conservation of Nature
MAB Movimento dos Atingidos por Barragens [Movement of Those Affected by Dams]
MAPE Movimento Arte e Pensamento Ecológico [Art and Ecological Thought Movement]
MDB Movimento Democrático Brasileiro [party]
MMA Ministêrio do Meio Ambiente [Ministry of the Environment]
National Front Frente Nacional de Ação Ecologica na Constituinte [National Front for Environmental Action in the Constituent Assembly]
OAB Ordem dos Advogados do Brasil [Brazilian Bar Association]
OIKOS União dos Defensores da Terra [Union of Defenders of the Earth]
ONG organização não-governamental [nongovernmental organization (NGO)]
PCB Partido Comunista Brasileiro [party]
PDS Partido Democrático Social [Democratic Social Party]
PFL Partido Frente Liberal [Party of the Liberal Front]
PMDB Partido Movimento Democrático Brasileiro [Party of the Brazilian Democratic Movement]
PNUD Programa das Nações Unidas para o Desenvolvimento [United Nations Development Program]
PPG-7 Pilot Program for Conservation of the Brazilian Rainforest (funded by G-7)
PrepCom (United Nations) Preparatory Conference
PROALCOOL Programa Nacional de Alcool
PROCONVE Programa de Controle de Poluição do Ar por Veículos Automotores [Program for Control of Air Pollution from Automobiles]
PSB Partido Socialista Brasileiro [party]
PSDB Partido Social Democrático Brasileiro [Party of Brazilian Social Democracy]
PT Partido dos Trabalhadores [Workers' Party]

PTB Partido Trabalhista Brasileiro [Brazilian Labor Party]

PV Partido Verde [Green Party]

RIMA Relatório de Impacto Ambiental [Environmental Impact Report]

SBF Sociedade Brasileira da Física [Brazilian Physics Society]

SBPC Sociedade Brasileiro para o Progresso da Ciência [Brazilian Society for the Advancement of Science]

SEMA Secretaria Especial do Meio Ambiente [Special Secretariat of the Environment]

SISNAMA Sistema Nacional do Meio Ambiente [National System for the Environment]

SIVAM Sistema de Vigilância da Amazônia [System for Surveillance of the Amazon]

SMA Secretaria de Estado do Meio Ambiente [State Environmental Secretariat (São Paulo)]

SOPREM Sociedade de Preservação dos Recursos Naturais e Culturais da Amazônia [Society for the Preservation of the Natural and Cultural Resources of the Amazon]

SOS Fundação SOS Mata Atlântica [SOS Atlantic Forest Foundation]

SPVEA Superintendência de Valorização Econômica da Amazônia [Agency for the Valorization of the Amazon]

SUDAM Superintendência do Desenvolvimento da Amazônia [Agency for Amazonian Development]

SUPREN Superintendência dos Recursos Naturais [Natural Resources Agency]

SUSAM Superintendência de Saneamento Ambiental [Environmental Sanitation Agency]

UNCED United Nations Conference on Environment and Development

UPAN União Protetora da Natureza [Union for Natural Environment Protection]

USP Universidade de São Paulo [University of São Paulo]

WWF World Wildlife Fund (also Worldwide Fund for Nature)

Introduction

In February 2005 Dorothy Stang, an American nun who had worked for twenty years among impoverished farm workers in the interior of the Amazonian state of Pará, was murdered on her way to a meeting with local activists about land reform. Her death, and the reaction to it, instantly recalled the murder in December 1988 of the rubber-tapper leader Francisco (Chico) Mendes in the western Amazonian state of Acre. In the wake of both, international observers and domestic activists decried the lawlessness in the region, and the government promised active pursuit of those responsible. In Acre the assassins had been sent by big ranchers, in Pará apparently by big loggers. Either way, it seemed that despite a decade and a half of environmental and human rights activism, multilateral investment, and federal efforts to engage state officials, gunslingers continued to rule the roost.

Although there is plenty of violent death in urban Brazil as well, more subtle forms of death also stalk their victims there. More than seventeen million people live in metropolitan São Paulo, roughly equal to the entire population of the Brazilian Amazon. For them, everyday activities like going to work can be as deadly as a gunshot—decades after problems were supposedly resolved. Public pressure forced the Clorogil factory in Cubatão, São Paulo, to close in 1978 after workers died from probable workplace contamination. Later, Cubatão was branded the "Valley of Death," with reverberations nationally and internationally. The state's first democratically elected government in seventeen years ordered a cleanup that began in 1982 and was deemed successful. However, in 2001 former workers of Clorogil and its sister company, the French multinational Rhodia, were still seeking compensation for their lingering health problems. As one strategy, they became the nucleus of a new Environmental Justice Network.

Tempting as it might be to say that nothing had really changed in the

Amazon in the seventeen years since Chico Mendes's death, that would be a mistake. Both gains and losses have been registered. Environmentalists now influence Amazon policy through the federal environmental ministry and in some other areas of the federal government. They are also involved in projects supported by multilateral lending agencies and northern governments, especially the Pilot Program for the Amazon, funded by the Group of 7 (G-7) of the world's wealthiest countries. In Amazonian states where environmentalism had already gained a foothold among those struggling for their livelihood, Acre and to a lesser extent Amapá, environmentalists achieved influence in state governments. Although the environmental ministry boasted—and rightly—that from August 2004 to July 2005 the rate of Amazonian deforestation fell by 31 percent (MMA 5 December 2005), the total area deforested each year of 18,900 km² is still greater than the average for the 1990s of 17,000 km² (*Notícias Socioambientais*, 6 December 2005). Illegal timber harvesting and smuggling remain highly profitable; and powerful land grabbers, hit men, landowners, and lumber companies continue to engage in illegal practices with impunity. At the same time, even as new legal instruments for conservation units and other protective mechanisms have been invented, criminality has become ever more pervasive in the political as well as civil society in Amazônia, derailing sporadic efforts by the federal government to assert control over affairs in the region. Nonetheless, approximately 67,432,419 hectares of land are now in 277 federal conservation units, and 30,176,431 hectares in state conservation units.[1] The increase in land placed in conservation regimes has been dramatic in the Amazon region in recent years, from a total of 24,933,170 hectares in 1989 to over 60,711,694 in 2006. As we shall see later, however, designation as a conservation unit is only one small step on the way to effective conservation of an area.

In urban Brazil the environmental picture is equally mixed. São Paulo state is Brazil's environmental policy innovator, with substantial environmental capacity in its environmental agencies and the largest concentration of environmental activists. In the 1980s activists teamed up with scientists and the state environmental agency to tackle air pollution, and brought fixed-point industrial sources of pollution to near global standards. Extreme episodes of air pollution have virtually disappeared, though their impact lingers in the damaged health of people and surrounding ecosystems. But now around 90 percent of air pollution comes from the 5.5 million cars that circulate in the megacity daily, a problem much harder to regulate than a few thousand smokestacks (CETESB 2000). In the 1990s São Paulo finally implemented a sanitation plan that succeeded in lowering the pollution load from domestic waste, having earlier instituted a program mandating treatment of industrial effluents. The Environmental Justice Network has built upon the accumulated know-how of activists in more densely organized parts of the country like São Paulo to keep companies from getting away with just moving polluting activities to less organized areas.

Many of the advances registered remain precarious, requiring constant monitoring and pressure from policy entrepreneurs and environmentalists inside and outside the state. This is because public policy decision making in Brazil is highly politicized (Rua 1997, 172), and there is rarely a last word. Enforcement tends to be weak, and expectations that policies will be enforced tend to be weak as well. As Levitsky and Murillo (2006) point out in their recent work on institutional weakness in Latin America, this weakness affects the way people behave in relation to political institutions and each other. These characteristics of the state strongly resemble those described by Douglas Chalmers in 1977 when referring to "the politicized state" in Latin America, a point to which we will return below.

These stories about the Amazon and São Paulo show the high stakes and daunting complexity of Brazilian environmental politics. Our first aim for this book is to describe this complexity as it has unfolded over time: Who has been able to shape Brazilian environmental politics, and by what means? How should we evaluate the resulting environmental politics and policy? We also look to explain some of the patterns we find. We begin by considering the relationship between international and domestic factors, arguing that domestic factors have been more important in shaping outcomes than often assumed. We also argue, though, that "domestic" and "international" are often intertwined in Brazilian environmental politics and that a framework focusing on the formation of networks in multi-level governance helps to make sense of the many interactions among levels of governance and kinds of actors.

International and Domestic Origins of Environmental Politics

Ideas about environmental protection have existed on the periphery of social thought and political agendas for centuries. Sustained and institutionalized attention to the environment is a much more recent phenomenon, taking hold only in the second half of the twentieth century. The widespread adoption of environmental protection measures over a comparatively short period has prompted scholars to look for explanations at the international level. Many of these explanations fit into the broad family of theories of international norm diffusion (Finnemore and Sikkink 1998). In these, the focus is on how norms are collectively generated or constructed in the international arena and then transmitted to domestic societies. Theories of norm construction and diffusion vary in their portrayals of how conflictual this process is.

At the less conflictual end of the spectrum, writers focus on the development of normative consensus that they assume can be assimilated straightforwardly in domestic settings. For environmental norms, one notable example of this kind of work is the statistical study of the emergence of domestic environmental protection carried out by Frank, Hironaka, and Schofer. As they state, "Our main arguments are thus that blueprints for the nation-state are drawn in world society, that such blueprints have, over time, increasingly specified environmental protection as a basic purpose of the nation-state, and that the provisions of such blueprints diffuse from world society to individual countries" (2000, 102). Their event history analysis does show strong statistical support for a relationship between emerging global environmental protection norms and domestic environmental protection practices around the world.

At this very aggregated and consensual level, however, the spread of environmental protection is measured in ways that flatten most of the real content of environmental politics in a country like Brazil. For those seeking parsimonious, macro-level theories, this flattening might even appear positive—Brazil would be one data point in support of their argument that national environmental organization accelerated after the 1972 Stockholm conference, as Brazilians created their first environmental secretariat in 1973.

But by failing to explore the actual mechanisms by which such processes take root, rather than simply occur, we would have no way of determining which processes were likely to prove robust and which ephemeral. The tiny national environmental agency established in 1973 (see chapter 1) was only important insofar as it initiated a very slow and gradual process of building on this first effort until twenty-five years later, Brazil had a national environmental agency with real capacity. It was not inevitable that this should happen. It is worth noting that like Brazil, neighboring Argentina created its first environmental agency in 1973, only to dismantle it several years later, and then finally start again in 1991 (Hochstetler 2003). Such experiences belie any simple and linear vision of how international norms are diffused, and we argue that the reason why norm diffusion is not such a straightforward process lies in domestic politics.

Other theorists do pay attention to the struggles involved in norm construction. Critical theorists generally see norm creation as a kind of power politics, in which ideas and meanings are themselves venues for political contention: "a critical perspective is one that questions our understandings of the world around us, particularly those we take for granted, in order to identify who is served by them and who is marginalized" (Stevis and Assetto 2001, 2). Some critical theorists question whether environmental norms are genuinely universal at all (Pasha and Blaney 1998, 436), while according to others the particular versions of environmental aims that gain hegemony reflect the interests of politically dominant actors, primarily in the global North (Middleton, O'Keefe, and Moyo 1993; Najam 2005). But because these theorists fix their attention on dominant actors and discourses, they sometimes miss the ways southern countries like Brazil have helped to set the terms of the debate, albeit not exactly as they would choose. Norms spread, but by virtue of their spreading they invite new actors into the debate over how the norms can be reshaped so as to become more inclusive and universal —a move that requires that they also become embedded domestically. More nuanced discussions of global struggles over norms help us to identify processes at work that stimulate and shape *national* debates over those norms, which in turn influence the country's global role. As nation-state representatives struggle over emerging international norms with others (and with actors not defined by nation-states as well), their positions are often shaped by audiences and adversaries at home, at least as much as those abroad, in a manner akin to, but broader than, Robert Putnam's two-level games model (Putnam 1988).

Our own past work has focused on several mechanisms of international influence on domestic environmental politics, both conflictual and not. By employing a boomerang strategy, transnational advocacy networks can sometimes bring together a variety of actors in issue campaigns to influence governments unwilling to respond to demands of their citizens (Keck and Sikkink 1998a, 12–13). These networks have expanded environmental protections, defended human rights, and achieved other collective ends around the world. International institutions like the United Nations have provided frameworks and focal points for discussing environmental issues in global conferences that generate numerous international plans and agreements (Clark, Friedman, and Hochstetler 1998; Friedman, Hochstetler, and Clark 2005). These conferences have been accompanied by abrupt increases in the number of domestic environmental protection measures taken (Frank, Hironaka, and Schofer 2000).

The best-known stories of Brazil in these contexts have stressed its resistance to environmental protection when pushed from abroad. In 1972 in Stockholm, the Brazilian delegation vigorously defended the right of developing countries to use their natural resources as they saw fit (Campbell 1973; Castro 1972; Guimarães 1991, 147–57). The occasion, the United Nations Conference on the Human Environment, followed upon the end of the first UN Development Decade and the beginning of the second; Brazil's position reflected the assertion by developing countries of an international agenda of their own (Krasner 1985). A generation later, when Amazonian deforestation captured the world's attention and a variety of international actors mobilized to respond, many Brazilians—in the government and not—responded with nationalist resistance (Hurrell 1992; Keck and Sikkink 1998a; Kolk 1996). Even more recently Cristovam Buarque, a senator, former governor of the Federal District, former minister of education, and former Workers' Party (PT) intellectual, took on a young American questioner in New York, who asked him to respond as a humanist and not as a Brazilian about what he thought about internationalizing Amazônia. He responded that as a humanist he would be willing to defend the internationalization of the world—its petroleum reserves, its treatment of children, its financial wealth, its great art museums, and other wonders. But "as long as the world treats me as a Brazilian, I will fight to make sure that Amazônia is ours. Ours alone" (*O Globo*, 10 October 2000). His statement, reproduced in countless e-mails and on at least 51,200 web sites as of this writing, had an extraordinary resonance in Brazil.[2] Stories like this one tend to create an image of Brazil as a late adopter of ideas about environmental responsibility, thus likely to be a receptor country in the process of transnational norm diffusion.

Yet even in 1972, when Brazilian diplomats were forceful exponents of pro-development arguments at Stockholm, environmentalism had already developed roots in Brazil. Environmental ideas have a very long trajectory in Brazilian cultural history (Pádua 2002). Scientists and nature lovers began to form conservation organizations in the 1950s, and in a survey in 2002, seventeen of the Brazilian associations registered as dedicated to the environment and animal protection had been created before 1970 (IBGE 2004, table 16). The "new environmentalism" arrived in Brazil during the 1970s as it did in the industrialized North. By then the states of São Paulo and Rio de Janeiro had already established technical agencies for pollution control. A national environmental agency was created in 1973 and gradually expanded its personnel and range. In the decades since Stockholm, the desire to protect the environment has coexisted with the active pursuit of economic development, making Brazil one of the few developing countries where both ambitions have strong and articulate defenders.

Our book aims to tell the part of the story of Brazilian environmental politics that the transnationalized narrative omits. We combine process tracing with a focus on key events, moving between individual and institutional levels of analysis and paying attention to motivating ideas and their fit—or lack of fit—with other contending ideas in their time. Generally speaking, international and transnational actors influence Brazil's environmental policy by engaging Brazilians who have the authority, charisma, or organization to bring about changes in policy and practice. Who are these Brazilians? What are the *projects* to which they are committed, the *trajectories* they have followed? What links them, if anything? Which domestic institutional and political conditions sustain or constrain them?

International Influences

Political developments in the period that we discuss contributed to and were affected by regional and global trends. They signaled the end of a period when assertive military governments in the Latin American region adopted ambitious missions to purge their societies of subversive forces while setting their nations on the path to greater power in the world. At the same time, the successor governments were constrained by the changes in the global political economy that began in the 1970s, from the transformation of international finance following Nixon's decision to renounce the gold standard, to the oil shocks of 1973 and 1978, to the rush of international bank lending that ensued, and the debt crisis caused by overenthusiastic borrowing after that. The great confidence in developmentalism (Sikkink 1991) that Brazilians expressed at Stockholm had waned by 1982, when Brazil was forced to go to the International Monetary Fund for the first in a long series of stabilization loans. Over the next decades, Brazilian economic policy was gradually reoriented away from development policies with a strong state role toward a more liberal market economy with much of the investment initiative in the private sector (Lopes 1996; Smith and Korzeniewicz 1997). Lowering tariff barriers lowered the price of imports, making it essential to increase exports to balance the books. In response, a massive expansion of export agriculture catapulted Brazil into the first tier of producers of soy, along with cotton, citrus, and other crops. The export crop frontier in turn pushed small producers off the land, leading them to seek new opportunities elsewhere and opening up hitherto unexploited parts of the savannah and rainforest.

International forces have influenced Brazilian environmental outcomes in contradictory ways. At the same time as foreign environmentalists and political leaders urged Brazilian authorities to protect rainforests and biodiversity in the Amazon, international financial institutions demanded a reduction in state responsibilities and personnel, and illegal drug and timber traffickers undercut state authority altogether. Thus international actors clearly matter in Brazilian environmental politics, but on all sides of the question. Their activities strengthen or weaken the resolve and resources of different sets of domestic actors, as they seek to advance or block proposals or generate new ones.

Particular physical attributes of environmental problems often make it possible to address them at many different levels and in different venues, some of which are international. Activists must decide at which scale to act and in which venues, insofar as their networks and resources give them an opportunity to do so. Over the time period that we cover, there has been a marked expansion in the scale at which Brazilian environmentalists can act and the choices and resources available to them as a result. Some appear regularly abroad or at international events, part of a group that Tarrow has recently called "rooted cosmopolitans" (Tarrow 2005, 28–29) and Steinberg calls "bilateral activists" (Steinberg 2001). Some foreign activists have become what Chalmers called internationalized domestic actors (1993), moving into Brazilian domestic political space for extended periods; others remain distant but steady allies. Still others appear in the story once or twice, and go on to other things. The zone of relational cosmopolitanism is suffused with power, but also dynamic, in that it is a zone of constant frictions resolved in varying ways (Tsing 2005).

Nonetheless, it is important to remember that for most citizens of most countries, the word for international is "foreign." "Foreign" intervention is always intrusive. How it is mediated makes a difference, and it can help when the mediators are domestic actors who are also participants in international society. We can identify five mechanisms of interaction between domestic and foreign actors that emerge in our story of Brazilian environmentalism: diffusion, persuasion, leverage, payoffs, and coercion. Cosmopolitans are crucial actors in these processes. They are the central agents in diffusion, the most likely to be aware of new ideas and models available elsewhere, and the most likely to try to import the ones that appear promising. Efforts at persuasion are more likely to be influential when mediated by people familiar with the variety of cultural languages in the conversation, facilitating the process of translation (Tsing 1997). Alliances with likeminded others abroad may help raise the salience of an issue, and in any case strengthen the will and capacity of the central domestic actors. Leverage requires knowing how vulnerable target actors are and what resources can be brought to bear to influence them—as well as the leveraging party's degree of commitment. Payoffs in the form of material incentives—that is, offers of assistance that are hard to refuse (sometimes outright bribes)—and coercion are both mechanisms by which actors without the ability to persuade attempt to use money or force (either physical or not) as a substitute.

Whatever the pattern these relationships took in Brazil, they look different through the lens of domestic politics than they do from abroad. Some of the reasons for this are obvious but worth repeating, because when we move into the relatively abstract language of actors, encounters, and so forth, we forget how important the particularities of places and people may be, and the ways they affect international relations. Brazil is a very big country—similar in size to the continental United States. It is by far the largest and most powerful country in its region. Brazil is extremely diverse—ecologically, socially, racially, culturally. Like people in the United States, Brazilians tend to think of their country as *sui generis*, and are much more interested in domestic affairs than in events in other countries; Brazilian "exceptionalism" is as strong in its way as American exceptionalism is in the United States. The Latin American region still has a limited political reality for Brazilians, despite the existence of Mercosur, the free trade area that Brazil forms with its neighbors. Therefore norm diffusion requires a more active process of encounter, in which Brazil's distinctive character is recognized at the same time as its participation is sought in constructions of international normative consensus.

The Origins of Environmental Debates in Brazil

In the field of comparative environmental politics, concluding that environmental policy bears heavy traces of other features of domestic politics is common in both early classic works (Enloe 1975; Vogel 1986) and more recent ones (Adeel 2003; Schreurs 2002; Szarka 2002). Like the authors of these works, we argue that Brazilian environmentalism acquired distinctive features from its domestic context: the problems that it faced, the institutional setting, and the timing—that is, the other simultaneous events and social processes. Three features of Brazilian politics are especially important for understanding its distinct characteristics. These are the development of environmentalism in the context of a democratizing transition from military to civilian rule, the impact of federalism, and the continuous interplay of the formal and the informal. Its emergence during the transition period helped to shape an environmentalism that is more politicized and further to the left than one sees elsewhere, what Brazilians call socio-environmentalism. This political context contributed to unusually strong interpersonal relations among environmentalists in state and civil society institutions, who work together in both blocking and enabling networks.

Democratization

Over the last decades of academic writing about Latin America, political transition and then consolidation have been major orienting concepts, generating a voluminous literature about the causes, dynamics, and consequences of the change from military to civilian rule. While this literature is too large and diverse to be easily summarized, several general conclusions have emerged that we adopt as starting points. One is that the transition was a "disjunctive" and incomplete process (Agüero and Stark eds. 1997, i). Some changes, like restoring elections, could be effected quickly through legislation, while others, such as creating mechanisms of political accountability or a fair and open judiciary, required more protracted efforts that are still incomplete. A second observation is that the near simultaneous transition to more market-oriented economies added another set of cross-cutting challenges. Brazil's economic transition was one of the latest in the region, coming in the 1990s after significant political changes had already occurred (Friedman and Hochstetler 2002). Neither the earlier developmentalist nor the later market policies challenged the profound economic inequities that are at global extremes in Brazil. While full discussion of these broader changes is beyond the scope of this book, their impact is evident in the specific stories of Brazilian environmental struggles.

Democratic political theorists have considered the relationship between democracy, democratization, and the environment. Their most common assertion is that democracy and related concepts like participation and decentralization are associated in positive ways with environmental protection (Doherty and De Geus eds. 1996; Press 1994). A large quantitative study failed to find a statistical relationship (Midlarsky 1998), however, and as Desai points out, the firmest conclusion may be that authoritarian regimes are usually unfriendly to the environment (Desai 1998, 10). Brazil's military regime did begin formal environmental protections, but in chapters 1–3 we show how the gradual transition changed the strategies of both state and societal actors.

Brazil's military regime, which took power in 1964, lasted longer than any of the other contemporaneous authoritarian regimes in South America. The slow transition from military to civilian government, lasting from 1974 to 1989, shaped the political opportunity structure of Brazilian environmental politics, implanting changes outside the environmental domain that became quite important within it. Sometimes these were events, like the amnesty in 1979 that prompted the return of activists with new ideas and strategies based on their experiences in exile—such as creating a Green Party. Sometimes they were legal changes leading to new tools that could be used by environmentalists and their opponents, such as the law to protect diffuse interests in 1985 or the participation-oriented constitution of 1988. As Brazil turned to more open electoral politics, both state and societal actors had to reconsider their political strategies and interests. Democratization clearly transformed Brazilian environmental politics, although not in a unidirectional or unilinear way.

State efforts to combat pollution expanded in the mid-1970s, as did environmental organizations. By the beginning of the next decade, the military's limited liberalization began to turn toward a full-scale democratic transition (Alves 1985; Stepan, ed. 1989). Meanwhile, myriad social movements mobilized to demand decent social conditions and a share of the material progress that recent high growth rates had produced, but whose benefits had gone disproportionately to the wealthy. The dramatic contrast between extremes of wealth and poverty was something that environmentalists could not ignore. Finally, democratization produced pressure for a wider distribution of power and decision making, both within state institutions and between these as a whole and societal organizations. Eventually framed as demands for a "new citizenship," these were political claims in whose construction environmentalists fully participated (Hochstetler 1997, 2000).

Environmental organizations in Brazil were forming and beginning to mobilize support within this large and growing multi-organizational field. Being an environmental activist did not preclude participating in party organizations, rebuilding student organizations, raising money to support union members on strike, or protesting the high cost of living. Indeed many of the activists who came of age politically during this period engaged in all these activities, experiences that strongly shaped what we call the second and third waves of Brazilian environmentalism. Democratization convinced them they needed to broaden the social bases of their appeal to be able to influence newly democratic decision-making processes in a context of economic crisis. Many of these activists continue to be central to environmental politics in Brazil, as often in the state as in society.

Based on this experience, many environmental mobilizations in Brazil have focused as much on social equity and participation as on protecting the environment. Brazilians call these preoccupations "socio-environmentalism," a homegrown set of ideas that emerged more or less simultaneously out of struggles in Brazil, India, Indonesia, and a few other developing countries. Analogues now exist all over the world, diffused by many processes of discussion and joint mobilization among global environmentalists, including the advocacy networks and UN conferences mentioned above. The more general form of what Brazilians call socio-environmentalism is well-captured by the "social greens" worldview discussed by Clapp and Dauvergne (2005, 11–16). Socio-environmentalism is an attempt to make compatible the struggles for environmental sustainability and for sustainable livelihoods. Opposed to a purely expansionist capitalism on social *and* ecological grounds, it argues that empowering poor people and responding to their demands for social equity must be an integral part of any solution to environmental problems (Viana, Silva, and Diniz, eds. 2001). In a sense, we could call it a sustainable development for poor people. A socio-environmental definition of the environmental agenda was never fully consensual among environmental organizations, but it was an important foundation of political networking strategies in the Brazilian sociopolitical context.

Federalism

Federalism is an enduring feature of the Brazilian political system, as it has been legally divided into multiple jurisdictions since independence—in all of Brazil's eight constitutions. Brazil's federal system, comprising twenty-six states and a federal district, has three constitutionally designated levels of political authority: the federation (variously called the nation, the union, the State), states, and municipalities (counties). Each normally has an elected executive and a legislative branch.

The twenty-seven federal units differ dramatically in physical and political ecologies. Brazil's territory includes the moist tropical forests of the

Amazon and the Atlantic Forest, large areas of savannah and the floodplain or wetland of the Pantanal, temperate forest and pampas in the South, the semi-arid Northeast, and a great diversity of coastal ecosystems. The Atlantic Forest (of which only about 7 percent of the original area remains) and the savannah, suffering from mercury pollution caused by gold mining, have been the hardest hit by human activities. The human geography varies as well. About 220,000 people make their home in the Amazonian state of Roraima, while similarly-sized São Paulo has over 31 million people. São Paulo's per capita income is nearly five times that of Maranhão. Infant mortality ranges from 22.32 per 1,000 live births in Rio Grande do Sul to 112.97 in Alagoas (1991 figures from PNUD 1998). In all cases, the higher levels of development are found in states in Brazil's South and Southeast regions, the lowest in the North (Amazon) and Northeast, with the Center-west region in between.

While the three-part administrative division is enduring, the division of power has not been. Even formally, Brazil has seen regular oscillations between more centralizing and more decentralizing dynamics to its federal arrangements. The recent democratization, embodied in the 1988 constitution shifted many responsibilities and resources to the state and municipal levels (Martínez-Lara 1996; Souza 1997), following upon a military government that had concentrated powers at the center. As Souza notes, this redistribution of powers has increased the numbers of actors who can veto policies as well as the amount of political conflict over them (Souza 1997, 10–11). From a different standpoint, decentralization allows input from more societal actors and can be seen as democratizing for that reason.

Chapter 6 (Article 225) of the 1988 constitution is specifically about environmental rights and responsibilities, and numerous other references are scattered throughout the document. In many areas of environmental and natural resource management, the three levels of government share responsibilities (Farias 1999). However, instead of dividing jurisdictions in advance, the constitutional text assigns national standards and plans to the federal government, questions of "local interest" to the municipal government, and the (small) remainder to states. Not surprisingly, these vague mandates have left a great deal to be sorted out in practice, often informally; even within the same policy area, procedures may differ from one state to another. Complicating jurisdictional issues even further, states can create additional levels of governance such as metropolitan regions or micro-regions when they think they are warranted (and can defend them against other levels of government).

Federalism allows considerable variation in subnational environmental politics and policies. Patterns of political behavior, partisanship, alliances, attitudes toward the state and the private sector, and social organization, as well as levels of development and sources of wealth and its distribution, all vary considerably from one state to another and from one municipality to another. These variations mean that approaches to environmental concerns differ widely, as does the capacity to deal with them (Ames and Keck 1997–98). A study by the Brazilian Association of Environmental Agencies (ABEMA) catalogued the wide disparities in resources available for environmental protection in the country's five geographic regions. The vast Amazon region had only 130 state forest police officers in 1991 (all in Rondônia state), for example, while the Southeast region had 1,405 in just Minas Gerais and São Paulo states (ABEMA 1993, 11). The same two regions had 25 and 119 personal computers available (ABEMA 1993, 24). We explore sources of subnational capacity and political alliances in the politics of natural resources in Amazonia and air pollution in São Paulo in chapters 4 and 5.

Although understanding federalism's impact on environmental politics is crucial, no simple "layer-cake" model will suffice (Grodzins 2000, 55). The three levels of government are carved up quite differently for various areas of activity and regions of the country, and both legal regimes and informal political relations affect the distribution of responsibilities. Consider water pollution. Pollution control is a state responsibility, but land-use zoning, a key component of an aggressive approach to controlling water pollution, is under municipal authority. Furthermore, if water pollution affects a municipality downstream that is in a different state, the other state and also federal authority come into the picture (Keck 2002). Thus both environmental disputes and environmental capacity have to be mapped onto a dynamic field of interactions that extends well beyond its immediate cause(s) and effect(s). Such interactions are frequently shaped by considerations that have nothing whatsoever to do with the environmental problem at issue—partisan considerations, inter-regional rivalries, experiences from past interactions, and so forth. In addition, while the international level is not the only important level for understanding the politics of the Brazilian environment, it is certainly one of the multiple levels that are relevant—but not always, and not always in the same way.

Because environmental concerns so frequently cross administrative jurisdictions, we find it useful to use a term that has been developed mainly in studies of the European Union, *multilevel governance* (Bache and Flinders eds. 2004). The diverse scholars who have explored this concept share the central observation that collective decision making has moved away from unitary central states and now takes place simultaneously on multiple territorial scales (Hooghe and Marks 2003). The "governance" component of the concept reminds us to look beyond formal political actors and institutions. Governance can be exercised by many kinds of actors, both state and nonstate. These may include international and domestic environmental nongovernmental organizations, the criminal networks that increasingly operate in the Brazilian Amazon, or hybrid actors of various kinds. Examining specific instances of multilevel governance requires that we look for both formal and informal linkages among units at different territorial levels, and different combinations of state and nonstate governance institutions. Hooghe and Marks want the concept of multilevel governance to reach beyond constitutionally stipulated nested jurisdictions to situations "in which the number of jurisdictions is potentially vast rather than limited, in which jurisdictions are not aligned on just a few levels but operate at numerous territorial scales, in which jurisdictions are task-specific rather than general-purpose, and where jurisdictions are intended to be flexible rather than durable" (Hooghe and Marks 2003, 237). Because these situations often take confounding forms that make it hard to distinguish among the constituent parts, the authors invoke as a metaphor M. C. Escher's lithograph of steps that both descend and ascend, apparently simultaneously. This second model better describes the mix of formal and informal that pertains in the Brazilian cases we studied.

Informal Politics in a Politicized State

The process of democratization and the federalist state organization have shaped the emergence and form of Brazilian environmental politics, but not always in predictable ways. Both democratization and federalism have always been ambiguous political projects: democratization has left pockets of authoritarianism as well as widespread inequality, while Brazilian federalism is full of vaguely defined jurisdictions, resolved through informal agreements. To some degree this arrangement stems from the extraordinary number of institutional reorganizations that have occurred over the last four decades (Martins 1997). More generally, Brazilian politics has a significant informal component to it that derives from the politicized nature of public-policy decision making.

Thirty years ago Chalmers noted that states vary on the degree to which institutionalized processes rule substantial aspects of their political systems. For the Brazilian state, descriptive aspects of the politicized rather than institutionalized state sound current: "the possibility of exerting effective influence outside established procedures means that the policymaking process is potentially created anew for each decision. . . . In an institutionalized regime there is a constant tendency to establish a fixed and recognized set of legitimate participants, set arenas for action, and rules for decision. In the politicized state, a premium is placed on redefining the groups, classes, and interests involved, the way in which they should encounter each other, and the way in which the outcome is determined" (Chalmers 1977, 25). Chalmers goes on to stress the importance of being "in power" to gain patronage and the authority to establish new programs and agencies, and change laws and decision-making procedures; he also points to the blurring of lines between administrative processes and party-electoral processes. One thinks, for example, of the more than twenty thousand federal jobs (*cargos de confiança*) under the control of the Brazilian president's office, and the comparably large number at state and municipal levels. In the United Kingdom, the number of jobs in the prime minister's gift is around a dozen.

Brazilian politics consequently plays havoc with expectations about path dependence, taking here Pierson's restricted definition of it as "social processes that exhibit positive feedback and thus generate branching patterns of historical development" (Pierson 2004, 21). The positive feedback (or self-reinforcing mechanism) in Brazil's case is precisely the nonresolution of key questions, the nondecisions (Crenson 1971), and the provisional nature of policy legislation, whose implementation requires a separate mobilization of commitment. Most Brazilian policy requires enabling legislation (*legis-*

lação complementar), in which the form of implementation is spelled out in exhaustive detail. This practice makes it easy for a legislator to vote in favor of a policy without ever intending that it be carried out; all that is necessary is to stall the subsequent enabling legislation, either by opposing it frontally or through legislative maneuvering. This, along with the regular changes in the structure and personnel of public bureaucracies, means that policy institutionalization is less "sticky" than institutionalist scholars might expect (Martins 1997). It depends substantially on the voluntarism of committed individuals, and owes as much to short-term improvisations (*jeitos*) as to a longer-term process of embedding procedures. Over time, paths are as likely to meander as to branch.

To understand the institutional pathways for policy making and political struggle, we therefore have to pay as much attention to informal channels as to formal ones. When we say that power often trumps institutional rules, we do not mean that institutions *never* work the way they are supposed to—often they do. However, because one can never be certain that they will, other kinds of political agency become crucial in examining a policy process. Rather than look at civil society pressures or at state-society networks as important just in the early stages of a policy process, we must look at them all the way through, from conception through enforcement, because the completion of one stage does not guarantee progression to the next. In some cases this is due to institutional weakness—instability over time and inability to enforce decisions—of the kind discussed by Levitsky and Murillo (2006). In other cases it reflects a lack of political will, or an inclination on the part of legislators or agency heads to decide something in a formal sense, while knowing that there is no possibility of enforcing it. However, in still other cases it reflects a lack of political capacity to articulate and mobilize the pieces necessary to make something happen. It is in this last kind of case that actors with strong interorganizational networks can make a significant difference (Abers and Keck forthcoming). Networks can provide to state and societal actors focused on a particular question the political, and sometimes even the technical capacity to produce action (Keck and Abers 2006). In this sense, the policy process requires a political mobilization all the way down. Even then, enforcement remains a difficult problem, as is evident in Brazil given the lamentable state of its civilian police apparatus (Arias 2006; Pereira 2006).

Networks and Environmental Politics

These complex, often informal, and multilevel relationships require social and political networks—made up of individuals as well as organizations. Networks play a central role in promoting activist agendas in Brazil. Networks connecting activists in civil society and committed individuals in state agencies facilitate issue advocacy and institutional change. The centrality of networks in constructing political capacity helps to explain how even in the absence of what political process scholars might identify as a "political opportunity" (Tarrow 1994), actors are sometimes able to identify and make use of more serendipitous "agitational niches" (Keck and Sikkink 1998b) to push their goals forward. The resources of networks can be mobilized either to block policies that they oppose (blocking activities) or to facilitate from a variety of angles and locations the adoption or implementation of policies that they support (enabling activities) (Keck 2002). Turning policy into practice demands this kind of enabling activity, given the weak enforcement capacity and low levels of institutional continuity characteristic of the Brazilian state. Each kind of activity frames policy questions in different ways, has its own time dynamic, emerges in particular kinds of political contexts, and favors the inclusion or exclusion of different kinds of actors (including international ones).

Blocking activities emerge in situations of threat to the environment. Networks frame environmental issues in adversarial ways, sharply dividing the actors who cause problems from those who would solve them. The sense of threat adds urgency to the time frame, even though actual mobilizations may stretch over years. Blocking activities take place in hostile political contexts, when unfavorable decisions are being made by public or private actors not easily influenced by the networks. On their own, the least powerful societal actors will have little recourse other than street protests to try to block policies. However, they will often find élite allies. State actors are frequently arrayed on both sides of the environmental conflict. Environmental sympathizers can act within the state to block bad policies or to drag their feet on implementation. State-based allies can support societal actors by slipping documents and other kinds of information to them that strengthen their protests. For obvious reasons, much of this kind of blocking activity is meant to remain secret, but participants have confirmed that it was common, especially during the military government but also afterward. Well-connected and well-informed international allies also are often important. Network members with access to national and international media outlets are especially useful, and efforts to influence public opinion are central to the strategy. Keck's and Sikkink's boomerang of international influence (1998a, 13) is a classic version of how these kinds of actors may come together in a blocking network, although it is not the only possible kind.

In contrast to the pattern in blocking activities, state actors will often be the coordinating participants in enabling activities, directly or again behind the scenes. These more cooperative arrangements frame environmental problems as having potentially consensual solutions that can be worked out through dialogue and participation, although there is likely to be conflict along the way. Finding these solutions often requires participants to approach problems through a longer time frame, both for working out the solution and for seeing the actual benefits of cooperation. This process resembles what Elinor Ostrom (1997) calls co-production, and Peter Evans (1997a, 1997b) calls state-society synergy. Cooperation must be made routine to sustain the longer time frame. These characteristics mean that enabling activities rarely have the visibility of blocking ones—another routine meeting is unlikely to grab headlines. Networks engaging in enabling activities often prefer to work in this private way, as public confrontation might prevent cooperation. International actors in particular tend to seek a low profile in enabling activities, as these actors are not long-term domestic stakeholders.

When we talk about networks here we are referring both to the kinds of social networks that arise from shared experiences or affiliations, and to networks that have been deliberately created as political or policy networks. Brazil's organizational ecology features a small number of large professional NGOS coexisting with myriad small local organizations dedicated to specific kinds of activity. A history of national and regional encounters, discussed in chapter 3, has helped to create links among them, and has made the network a modular form of organizational design. We therefore see a climate change network, the Atlantic Forest network, an environmental justice network, the Amazon Working Group, and many other networks. In 1994 a survey found that 67 percent of Brazilian environmental organizations reported participating in at least one network (Pizzi 1995, 40), a pattern that continues today. Even without a formal network analysis, it is possible to develop a fairly clear idea—from responses to interview questions, jointly signed documents, co-sponsored campaigns, and the like—of which organizations work together most often and in what areas.

Our observations about personal connections important at key moments are drawn from statements by those involved and also from personal observation during close to fifteen years of attending meetings and events in Brazil during visits made at least annually. Because we have seen commitments and trust among people develop over time, we have become interested in the trajectories of those who recur in the stories we tell. Most of those who remain active in environmental politics over a number of years occupy a wide array of roles during that time, as a result of professional development, developments in their personal lives, or political opportunities. The interaction between individual trajectories—during which relationships, trust (or mistrust), and evaluations are built—and institutions becomes the privileged site for examining political action in Brazil. Specifically, the reputations and relationships that people bring with them into political appointments constitute the social capital on which they must be able to draw to activate successfully the institutions whose actions they will try to shape.

Our task is to explore how state and nonstate actors, international and domestic, come together to form the multilevel governance of the Brazilian environment. We begin with three chapters that trace the development of state and societal environmentalism from 1972 to 1992. Chapter 1 isolates formal, national-level state actors for (temporarily) separate analysis, discussing the specifically environmental legislation and agencies that were created in the 1970s and 1980s. These are a linchpin of Brazil's environmental policy capacity, and their development helps shape both subnational environmental politics and Brazil's interaction with the international arena. The chapter also explores how the transition from military back to civilian rule transformed the ground rules of Brazilian politics during the 1980s, including environmental politics.

Chapter 2 focuses on the first conservationist wave of Brazilian environmental activism, which emerged in a fairly hostile environment of pro-developmentalism, in both civilian and then military government. It also discusses the second wave of more politicized environmental movements that responded to the democratizing politics of the political transition. Chapter 3 then traces how environmentalism was further transformed during the 1980s and 1990s in response to both democratization and the entrance of transnational actors. Environmentalists responded by professionalizing their organizations and by adopting a discourse of socio-environmentalism. Collectively, these first three chapters demonstrate that the development of Brazilian environmental protections began well before the surge of international attention to Brazil's natural resources in the 1980s and early 1990s.

Even in these early years, however, separations between state and society and between national and international are somewhat artificial. Early foundations for relationships between state and nonstate actors were laid during the 1980s in particular, some of them individual and others increasingly institutionalized. The links developed even more rapidly after the democratic transition coincided with greater international attention in the late 1980s. We take account of these networks with more holistic views of the politics of the Amazon in chapter 4 and of anti-pollution efforts in chapter 5. Disputes over the politics of Amazon protection were significantly internationalized after the mid-1980s, at both governmental and nongovernmental levels. Anti-pollution efforts, by contrast, continue to present a story that relies much more on the links between domestic state and nonstate actors at multiple levels of the federal system, with comparatively little attention from international actors. Our Conclusion reviews the different and contradictory roles that international actors have played in Brazilian environmental politics, and examines patterns in the relationships among international, national, state, and local politics.

Building Environmental Institutions: National Environmental Politics and Policy

In this chapter we introduce key components of national environmental institutions as they were constructed from the 1970s to the present, in roughly chronological order. Institutional innovation took place against a variety of political backdrops: the military regime that governed Brazil from 1964 to 1985; the political opening that began in the mid-1970s and continued through the promulgation of a new constitution in 1988; challenges to authoritarian structures and patterns of decision making dating back much further than 1964; and the inherent difficulty of assuming new public responsibilities in a time of economic crisis, tight budgets, and diminished state capacity. Environmental institutions introduced during the military regime had to find creative ways to circumvent the opposition of powerful developmentalist currents in both state and society. During the democratic opening, environmental movements demanded not only new policies but also new venues in which they could participate. They were not alone—other civil society organizations that emerged during the transition mistrusted the state's ability (or desire) to look out for the public interest, to say nothing of its capacity to do so.

The development of state environmental institutions thus took place in a period of considerable political upheaval, from which it was by no means insulated. Environmentalism itself became a lightning rod for debate over whose interests were being defended, and for what purposes, domestic or foreign. At certain moments, as during the furor in 1988–89 over high rates of deforestation in the Amazon, institutional changes were at least partly motivated by a desire to placate foreign interests. Nonetheless, where domestic actors were ready to take advantage of these openings, they did not remain changes made purely for foreign consumption, or as Brazilians would say, *"para ingles ver"* (for the English to see). Although international considerations sometimes enter the picture, the national process that we portray here was primarily structured by domestic concerns. Similar interactions took place in many state and municipal governments, and in new collaborative organs formed at both these levels. We particularly focus on how institutional changes at the national level were affected by and in turn affected the relationship between state institutions and societal actors, by exploring in greater detail two key moments of change: the National Constituent Assembly that wrote Brazil's 8th Constitution, promulgated in 1988, and the process set in motion by passage of a law to protect diffuse public interests, in 1985.

Brazil's military government has a deservedly poor reputation for its treatment of many environmental matters (Guimarães 1991; Mahar 1989; Zirker and Henberg 1994). Still, unlike most of its counterparts in the region, it did begin creating a regulatory and institutional framework for environmental protection. The regime issued nineteen federal laws, or decree laws, and twenty decrees on the environment during its years in power (compiled from Republic of Brazil 1991), besides establishing a national environmental agency in 1973 and a national system of environmental policy in 1981. By contrast, the Argentina junta that took power in 1976 dismantled an existing environmental agency (Hochstetler 2003), and the Chilean military government's ideological bias against a large state role restricted its environmental capacity building (Silva 1997).

Although environmental policy and agencies have a much longer history in Brazil than in its neighbors, early environmental capacities fit with the peculiar politics of military government. The regime's agenda embraced environmental policies aimed at rationalizing resource use, such as data collection on the physical environment or establishment of standards (Monteiro 1981, 30–31), but environmental regulations could not challenge the military's security and developmentalist priorities, nor could they draw on broader participation. To get around this obstacle, Brazil's first national environmental secretary, Paulo Nogueira Neto, employed what he calls "environmental guerilla activities" and relied on the authority of his reputation as a scientist rather than seeking support from fellow environmentalists. This institutional framework was remade several times after the military left power (see table 1.1 and discussion below). Besides increasing the size of environmental agencies, governments during the transition made rhetorical commitments to democratic participation in decision making that environmentalists struggled to make real in practice. As in other domains of the transition, opposition elites and impatient citizens seized small openings and enlarged them further still (Alves 1985).

Success in obtaining legislation in Brazil was no guarantee that it would be enforced, or even enacted. Most ordinary legislation requires further, complementary legislation to establish procedures for implementation—something that elsewhere frequently occurs within the administrative bureaucracy. Even then, the power to monitor compliance and sanction violations may be lodged in one or more other agencies, or nowhere at all. When Brazilian environmentalists fight for new legislation to curb environmental abuses, they know that its enforcement will often require an entirely new struggle. Moreover, democracies differ a great deal in legal culture and in assumptions about the law's application. In Brazil the law's application is widely understood to be socially contingent (McAllister 2004). Brazilian elites expect special treatment; the less powerful expect to be victimized rather than protected by the police. For the powerful, there has always been a *jeito*—a way around things, generally understood to involve a present or future exchange of favors, though not necessarily a bribe (Barbosa 1992; Da Matta 1987). The powerful expect the state to pay attention to their concerns; the less powerful do not. In this context, finding ways to operationalize environmental goals and make them stick is environmentalists' greatest challenge. Still, legislation and regulations, however formal, create legitimate instruments for trying to hold the state or firms accountable for their failure to act—a process in which an increasingly powerful Ministério Público (roughly translatable as Public Attorney's Office) would prove a crucial ally.[1]

Another source of unpredictability in Brazilian public administration is the appointments process. Most policy-making positions are filled by political appointment in Brazil, either directly by the chief executive or indirectly by his or her appointed minister or secretary. At the end of the 1980s the Brazilian president had personal appointments power over fifty thousand jobs, compared with several thousand for the president of the United States and several dozen for Japan's prime minister (Schneider 1991, 6). The figure for Brazil dropped to just over 21,000 jobs in 2005, three-quarters of which were filled by people who had in fact passed the civil service exam (*Brazil Forum*, 16–22 July 2005). Appointments structure decision making, and because elites build networks as they circulate through the bureaucracy over the course of a career, they structure interagency relations as well. Although someone appointed to a position of authority in an agency may be someone who has made a career within that agency, he or she could as easily be a complete outsider. Especially with technical agencies such as environmental agencies, the appointments process is central to explaining fluctuations in an agency's influence and attitudes over time.

Elites dependent upon political appointments generally spend part of their careers outside government as well. Just as some of them—what Schneider calls the "political *técnicos*" (1991, 6)—become brokers and intermediaries within and among agencies, they may also become brokers linking networks of public officials with pressure groups or social movements outside politics. Personal connections remain central to Brazilian bureaucratic politics, and the personal trajectories of key officials allow them to accumulate social capital upon which they can draw for later action. During the period that we examine, newly elected governors or mayors often named people with environmental experience to administrative positions, helping to maintain the fluidity of state-society relations noted in the Introduction. The appointments process presents some clear opportunities for political engagement, but also undermines the institutionalization of policies and practices in the bureaucracy.

Building Environmental Institutions in Brazil: The First Environmental Secretariat

Modern Brazilian environmental politics began in the 1960s, as the military regime actively consolidated legislation relevant to the environment (Monteiro 1981). Between 1965 and 1970 the forest, hunting, and mining codes were rewritten, and between 1965 and 1969 a sequence of laws was passed leading to the National Sanitation Policy (*Política Nacional de Saneamento*). In 1971 a special commission in the Chamber of Deputies held a three-day conference to discuss the problem of pollution, debating the establishment of a national pollution control program. By 1972 there were at least thirty-four public organs at the federal level, distributed among nine ministries and one secretariat of state, whose decisions had direct or indirect impacts on environmental policy.

Inspired by the Stockholm Conference, Brazil's military government took the next step of consolidating some of those responsibilities in a Special Secretariat of the Environment (SEMA) at the federal level. Only eleven countries had such an agency at the time. The new institution was charged with promoting conservation of the environment and rational use of natural resources. It was also to set national norms and standards with regard to pollution (Monteiro 1981, 30). Paulo Nogueiro Neto was named the first environmental secretary, a post he would hold for twelve years (Guimarães 1991; interviews with Nogueira Neto 1991, 1992). The secretariat began in 1973 with almost no budget and just three employees occupying two rooms in the Ministry of the Interior. It was ill equipped to challenge the much larger and wealthier bureaucracies administering the military regime's economic and security policies, often damaging to the environment. SEMA therefore defined its domain narrowly, and relied heavily on the informal strategies and networking of Nogueira Neto.

Nogueira Neto, a lawyer and biologist, had been an environmental activist in São Paulo since the mid-1950s. In 1955 he co-founded one of Brazil's first conservation associations, the Association in Defense of the Environment (Associação em Defesa do Meio Ambiente), after campaigning to support Governor Jânio Quadros's proposal for a protected forest reserve in the area of Pontal do Paranapanema, in the southwestern part of the state. Active in international conservation circles, he was named to the executive board of the International Union for the Conservation of Nature (IUCN) in 1970 (*IUCN Bulletin* 2 (1970), 14).

Nogueira Neto had few allies in the government. The economic and foreign affairs ministries were openly mistrustful, believing that environmental concerns could create obstacles for Brazilian development. He describes the reigning view as one according to which "Brazil was an island under siege by the rest of the world . . . that . . . had to defend itself" (interview with Nogueira Neto 1992). Itamaraty, the foreign ministry, believed that developed countries would make attention to the environment into an instrument of imperialist domination. The economic agencies in government were similarly hostile, and Minister Delfim Neto continuously blocked resources for the agency.

However, personal networks helped. Nogueira Neto made friends with diplomats, who respected his status as a university professor and recognized his political lineage. The new secretary's great-grandfather had been a presidential advisor, and his father a federal deputy; he knew how to move in political circles. Relationships with the press were likewise good. Once again, personal networks were crucial. Nogueira Neto's father had been exiled under Vargas's regime alongside Júlio Mesquita, owner of the *Estado de São Paulo* media group, and he had spent holidays with Júlio Neto, the current owner. Rogério Marinho, brother of the head of the Globo media empire, Roberto Marinho, also became a good friend. So *Estado de São Paulo* and *Globo* were consistently supportive, as was the newspaper *Jornal do Brasil* more occasionally.

In the stories that Nogueira Neto tells about his tenure as environmental secretary, personal networks, publicity, and *jeito* were the source of the advances made. He tried to portray his position as a strictly technical one; his only possible source of authority was superior technical knowledge. Startled by the effectiveness with popular opinion of his claim of this knowledge—and how publicity and public opinion could move even a military government—he sought first to alert public opinion to the problem of polluted beaches, a problem that had long concerned him as a biologist. And convinced that people only become motivated when a problem is dramatized, he gave an interview in Santos, on the São Paulo coast, saying that marine pollution along its beaches was a potential source of hepatitis. Authorities rarely made such statements at the time, and that week the Santos beaches were virtually deserted. The mayor countered by calling upon journalists to witness him swimming at the city's beach, and the hotel keepers in Santos accused Nogueira Neto of trying to promote hotels in the interior of the state. But the result was that the state government resumed construction on underwater sewage pipelines, work that had been suspended for being too costly. The beach remained polluted, but much less so than before. The Santos case was the new secretary's first lesson in the power of public opinion. He subsequently repeated the experiment in Salvador, Bahia, where the state government asked him to help resolve the same problem. "Had I gone first to the government, they would have said, 'we'll look into it, we'll see . . .,'" he recalled. "But once it's in the newspaper . . ." (interview with Nogueira Neto 1992).

A conservationist at heart, Nogueira Neto lamented how ineffectively the current institutions were protecting forests, even those inside conservation units. "We engaged in a kind of environmental guerilla activity, one aspect of which was to occupy available space," he recounted. Since national parks and forests were under the auspices of the Brazilian Institute for Forestry Development (IBDF), then part of the Ministry of Agriculture, the new secretariat came up with a creative alternative:[2] ecological stations set up in parallel to the IBDF's parks, covering around 3.2 million hectares of land by the early 1990s. The term "station" (*estação*) evoked experimentation and research rather than conservation per se, so when the secretariat submitted a bill authorizing the creation of ecological stations, it passed unanimously. The law allowed 10 percent of the areas to be used or modified for research purposes. State governments ceded land, with permission for their universities to use it. INCRA, the federal land reform institute, allocated around 2 million hectares in the state of Amazonas, the navy provided an island on the Paraguai River of which the secretariat had not even been aware, and organs like the Funding Agency for Studies and Projects (FINEP) financed research.[3] Even agencies whose activities financed destruction, like SUDAM (Agency for Amazonian Development), contributed money. This gave the secretariat access to resources that IBDF could not get, despite its formal responsibility for protected areas.

Ecological stations had no budget line, and the secretariat was not allowed to hire guards. This was a problem, as SEMA had around 317 employees to the IBDF's six or seven thousand. Each ecological station had two to four resident caretakers, who called INCRA and the federal police whenever they detected an invasion. Nogueira Neto had had written into the law a stipulation that the land had to belong to the public sector, to avoid creating the kinds of "paper parks" that the IBDF created, with boundaries drawn around private land whose ownership was often unclear, and there was no money to buy it anyway (see also Foresta 1991). To resolve the personnel problem, Nogueira Neto invented the word "ecometrics," meaning environmental measurement, and contracted with the highly professional Brazilian Association of Sanitary Engineers to provide people to carry out "ecometric services" in the ecological stations. For fifteen minutes a day, these people would measure and record temperature and rainfall; the rest of the time they were guards. The environmental secretariat used this subterfuge for years, paid for with money budgeted to contract for studies. The ecometricians were finally designated public employees in the early 1990s.

International contacts intensified during the 1970s, especially through the networks of scientists and conservationists of which Nogueira Neto had long been part. In the summer of 1965 a biology graduate student, Thomas Lovejoy went from the United States to Belem, Pará, to study birds. The Belem-Brasília highway was under construction, and people were already remarking that deforestation was following the road. Lovejoy established ties with prominent Brazilian naturalists and research institutions, and returned to Belém for two more years. In 1973 he became a program administrator at the World Wildlife Fund (WWF), arguing that if WWF was going to worry about biodiversity and wildlife, it had to worry about tropical forests as well. He built the WWF tropical forests program, and in 1976 he returned to Brazil with other WWF directors to meet with Paulo Nogueira Neto. The same year, he made contact with Maria Teresa Padua from the IBDF at the IUCN's general assembly in Zaire. Subsequently Lovejoy visited Brazil several times a year and supported or brokered support for many projects. He persuaded the U.S. National Science Foundation to give money to the critical minimum size of ecosystems project of the Institute for Amazonian Research (INPA) (interview with Lovejoy 1992),[4] which later became a joint research endeavor with the Museum of Natural History of the Smithsonian Institution. He helped Nogueira Neto find funds to finance the ecological stations (interview with Nogueira Neto 1992). Lovejoy went on to become deputy secretary of the Smithsonian, chief biodiversity advisor for the World Bank in 1998, and then the bank's lead specialist for the environment for the Latin American region. In 1988 he was the first environmentalist to be decorated by the Brazilian government with the Order of Rio Branco.

Despite these gains SEMA's position within the military government remained too fragile for it to seek either to openly expand its mandate or reach out to organized civil society (interview with Gualda 1991). Putting environmental protection first already meant challenging the developmentalist economic model; the armed forces had made the linkage of development and national security a core value, and the justification for its own role (Stepan

1973). SEMA's employees feared that promoting active citizen participation in its initiatives would invite suspicion that the secretariat was "subverting order." Even educational initiatives were timid. In 1976, when the Ministry of Education mandated that ecology be taught in primary schools, it gave schools a way out by recognizing a lack of qualified personnel to do the teaching (*Estado de São Paulo*, 25 January 1976).

While SEMA was putting out small fires, government policies were starting big ones. The onset of cheap credit, as international banks sought to recycle petrodollars in international markets, promoted large-scale borrowing to finance big infrastructure and development projects. In the midst of the OPEC oil shocks, the military pushed for completion of the Transamazon Highway, intended to integrate the national territory. In the highway's wake came deforestation and cultural (sometimes physical) genocide of indigenous peoples (Browder and Godfrey 1997). The Carajás project opened up huge mining operations in southern Pará (Carvalho 2001); the giant Itaipú Dam was built; and eight power plants were planned as a result of the Nuclear Treaty of 1975 with Germany. The first of these, Angra, was built on a site whose instability provoked frequent shutdowns. Even apparently pro-environmental measures, like the program to produce ethanol from sugar to provide renewable energy for Brazilian cars, displaced food crops and produced highly toxic wastes as a by-product.[5] Although SEMA was a member of the government councils that approved these projects, it served mainly as a rubber stamp (Guimarães 1991, 120–25).

However limited its role in the 1970s, SEMA benefited from a continuity of mission and structure. Then the passage of a National Environmental Law in 1981 initiated a series of reorganizations of environmental institutions that continued into the mid-1990s. At issue in these reorganizations were four big questions: the division of authority between federal and state governments with regard to both environmental conservation and pollution control; how organizations outside the state would be incorporated into decision making; the autonomy of environmental institutions in relation to other federal organs; and the power of federal environmental institutions to affect the policies and behavior of other state institutions.

The Creation of SISNAMA and CONAMA: The Council Model

The legislation establishing a National System for the Environment (SISNAMA) in 1981 and its first National Environmental Policy activities in 1983 were a long time in coming, their trajectory affected by the political transition under way.[6] The SISNAMA legislation had been in the making since 1975, when a Consultative Council with representatives from various ministries and technical institutes wrote a text that included measures for education, prevention, and monitoring (*Folha de São Paulo*, 11 June 1975). Although SEMA sent a draft law to the interior minister, Rangel Reis, in October 1976, contention over implementation measures held up its passage for years. Especially controversial was the proposal to create a special environmental fund, financed by a tax on polluting or potentially polluting firms (*Estado de São Paulo*, 11 October 1977).

In mid-1981 the country's military president, Figueiredo, finally moved on several pieces of legislation, sending a mixed message to environmentalists. On 27 April he signed a law that allowed ecological stations to be created by federal, state, or municipal governments, and that created the new category of Environmental Protection Areas (Areas de Proteção Ambiental), authorizing human presence in the area while restricting human activity (Law 6902). On 2 June he signed the preliminary bill defining environmental policy and creating a National Environmental Council (CONAMA), but not another bill before him that would have provided for a new forest policy in the Amazon and had provoked substantial opposition (discussed in chapter 4).

Why then? Between 1977 and 1981 environmental activism had become much more widespread, especially in the South of Brazil, as repression of social movements eased. By 1980 the environment was becoming a political issue, linked to a critique of the developmentalist economic model (Monteiro 1981, 113). The administrative structure that SEMA was trying to set up reflected the shift during the transition to democracy toward decentralization and participation as new guiding principles.[7] State governments were authorized to shut down sources of pollution for fifteen days, something that had generally required explicit presidential agreement before. Fines for degrading the environment increased. These measures at least appeared to be granting more power to the states as part of the democratizing process. A more cynical observer would point out that the federal government also needed to liberate itself from some of its obligations in the context of economic crisis at the time (interview with Sarmento 1999).

In any case, SEMA was authorized to present a very decentralized draft law to Congress in 1981, and congressional amendments strengthened the law considerably. Brazil's military government had been unusual in allowing a partly elected Congress to exist, and these kinds of changes were typical of its growing assertiveness (Lamounier 1989a). Congress set up a mixed commission in which members of the government and opposition parties jointly proposed modifications that SEMA had been afraid to include in its own draft. To Congress goes the credit for introducing environmental licensing, environmental impact studies, and units for carrying them out, all of them presaged in the draft law. The commission approved twenty-four amendments, reinforcing the twenty-five original articles. Considerable authority devolved to the states, remaining with the federal government only if states did not establish environmental organs. Congress kept the new National Environmental Council, CONAMA, from the preliminary legislation, which included representatives from federal and state agencies and also nonstate actors. CONAMA's job would be to discuss policies and recommend environmental standards, and SEMA would respond. In its long trajectory from project to law, the SISNAMA did lose the fund that would have provided it with an autonomous financing stream.

The law, number 6938, was approved unanimously, but required enabling legislation to take effect. Nogueira Neto believed that it passed because nobody knew much about it. Had it been in the papers, there would quickly have been a lobby against it from both inside and outside government. On the inside there was rivalry with the IBDF, on whose terrain SEMA was encroaching. On the outside, the National Confederation of Industry (CNI) fiercely opposed provisions criminalizing certain polluting activities, and lobbied successfully against them. Viewing the bill as a serious threat to business interests, the CNI urged Figueiredo to veto thirteen of its twenty-five articles. Instead he vetoed only the two articles that Nogueiro Neto opposed (interview with Nogueira Neto 1991; McAllister 2004, 105–6).

The resulting SISNAMA established the first institutional framework within which environmental protection could be addressed holistically, from federal to local levels and across a variety of sectors. It introduced the idea, later included in the 1988 Constitution, that each level of government should control environmental matters specific to its own geographical scope. However, like the Constitution, it failed to specify which responsibilities fell to each level, while stressing the importance of policy coordination among them. One concrete mechanism for ensuring coordination was the presence on the CONAMA of representatives from state- and municipal-level environmental agencies. Representatives from state agencies have been especially strong voices in this norm-setting body, with a seat for each, while municipal bodies have been underrepresented (interview with Sarmento 1999). Another idea present in the SISNAMA legislation but only fully developed later was that the Ministério Público could be used to defend public interests in civil courts (McAllister 2004, 36).

The formation of CONAMA was an especially important step, as it was one of several standard-setting institutions originating toward the end of the military regime in the early 1980s. The idea behind these councils was that state and nonstate actors would work together to establish basic standards and parameters that would orient future policy making and implementation. As an environmental council, CONAMA set technical limits (such as allowable levels of specific pollutants in the air), as well as procedural standards (notably procedures for assessing environmental impact). CONAMA included industrial and labor groups along with scientists and environmental activists as nonstate participants.

The 1988 Constitution later enshrined councils of governmental and nongovernmental participants at all levels of politics as mechanisms to institutionalize social movements' hopes for greater participation in politics (Tatagiba 2002). Councils dealing with social issues like health, social assistance, and children were set up in the early 1990s and seen as important conquests for social movements (Draibe 1998). On the environment, CONAMA's role at the federal level was reproduced in state and municipal councils. New water basin management councils were set up from the mid-1990s with a mandate to promote water quality as well as establish priorities for its allocation among uses (Abers and Keck forthcoming). These institutions have varied in how well they work as mechanisms of interest representation (Friedman and Hochstetler 2002).

CONAMA itself was slow to get off the ground, meeting for the first time only in 1984 (and only once in that year). It did provide a small opening for citizen participation, in that three environmental organizations were allowed to participate in its policy discussions, but they were selected and their participation was monitored by the military government. Most environmentalists did not see this as much of a political opening, although there were more opportunities for participation in the environmental councils of some states, beginning with the election of governors in 1982 (Antuniassi 1989).

National Institutions in Transition after 1985

In the first half of the 1980s the political agenda was consumed by the final transition back to civilian government. Through more than a decade of organized opposition to the military government, civilian politicians had put off specific plans for the post-military years. The Democratic Alliance coalition that backed the election of Tancredo Neves, a conservative but prodemocratic opposition leader, was the last gasp of this effort to submerge differences (Martínez-Lara 1996). When Neves died before his inauguration, his vice president José Sarney, former head of the pro-military party, became the new chief executive. Sarney's political weakness and lack of a coherent vision exacerbated the inevitable factionalization of the political system after the transition from military to civilian rule. What was true for the political system in general held in the environmental arena as well.

On his inauguration day, 15 March 1985, Sarney extensively reorganized the national ministerial structure, moving SEMA and CONAMA from the Ministry of the Interior to a new Ministry for Urban Development and Environment. This move put environmental administration together with such responsibilities as housing, basic sanitation, and urban development. The secretary of the environment, Nogueira Neto, resigned in July 1986, alleging that the new ministerial home was indifferent to SEMA (Guimarães 1991, 200) — although other sources say that Nogueira Neto was pushed out for his ties to the military and his indifference to urban areas (Foresta 1991, 250). CONAMA remained as a forum for policy deliberation, contributing in turn to further institutional change. In 1986 CONAMA increased its size to over sixty members, adding five more representatives of the environmental movement, to be elected by environmental NGOs registered with CON-

AMA in each of five regions. Three other environmentalists would continue to be appointed by the national executive.

CONAMA also began to flex its normative and regulatory muscle in ways probably unanticipated by the military government. Most notably, in 1986 it began to require environmental impact assessments for all activities potentially degrading to the environment (Resolution 001/86). Reports on the assessments needed to be made publicly available and could be publicly discussed. The requirements held for state as well as private enterprises, an important provision given the weight of state enterprises in Brazil's pollution problems. Two years later the new Constitution elevated these requirements to constitutional status. Over time, the requirement for prospective environmental impact assessment has had mixed success in environmental terms. Notably, it has been more successfully implemented in states like São Paulo for urban pollution than in northern (Amazon) and northeastern states for other kinds of pollution (Glassen and Salvador 2000). In addition, weak oversight has meant that good environmental plans do not necessarily translate into good environmental practices. Nonetheless, one positive indicator of the stringency of environmental assessment requirements is that a network of pro-business actors expended considerable energy to modify the legislation in the 1990s.

The new arrangement lasted only three years. In 1989, pressured by foreign environmentalists and their governments to slow the massive deforestation of the Amazon rainforest, Sarney again relocated the national environmental administration, this time to a new agency, the Brazilian Institute of the Environment and Natural Resources (IBAMA).[8] IBAMA was briefly the lead environmental agency, reporting directly to the president, and included not only SEMA and CONAMA but also the agencies for fishing development, forest development (IBDF), and rubber. With the personnel from these three agencies, IBAMA was created with 6,230 employees, a dramatic increase in size since SEMA had begun with its three employees (McAllister 2004, 39).

Between 1987 and 1989 the Sarney administration designed a major program to build the institutions and activities of environmental protection. The World Bank and the German Development Agency KfW (Kreditanstalt für Wiedraufbau) gave more than $100 million in loans and donations to support these activities well into the 1990s (Barretto Filho 2003, 334). The funds facilitated regular reforms of the national environmental bureaucracy (see table 1.1). Upon taking office in 1990 President Fernando Collor de Mello restored SEMA to its position as the chief environmental agency, with IBAMA as its implementation arm. He also gave the secretariat a more permanent status, reporting directly to the president. Furthermore, Collor weakened CONAMA's role, relocating its functions to a new state-only Council of Government (Conselho de Governo) composed of twelve ministers, of whom three represented military ministries.

In a surprise move, Collor chose a noted environmentalist, José Lutzenberger, for the post of environmental secretary, buying himself a honeymoon from the barrage of foreign criticism about the rate of Amazonian deforestation. Stephan Schwartzman of the Environmental Defense Fund, one of the leading Amazon campaigners in the United States, told the *New York Times* that Lutzenberger's appointment was "stupefyingly positive" (*New York Times*, 6 March 1990). However, Lutzenberger disappointed those who expected his appointment to open up more space for environmentalists to participate in government at the national level, demonstrating more interest in working with foreign environmentalists than with Brazilian ones. He took criticisms of the Collor administration's environmental policy as personal insults, and indeed many prominent environmentalists remained silent for a long time rather than offend a former ally.[9]

The exception to Lutzenberger's exclusion of Brazilian activists was in the Amazon, discussed in chapter 4, where environmentalists were already working with foreign organizations to put pressure on the government from the inside and outside at the same time. During this period (1990–92) the environmental secretariat initiated its programs on traditional peoples, developing programs involving rubber tappers, indigenous peoples, and environmental activists. The Collor administration intended for these policies to play a role in foreign affairs as well as domestic policy; unlike his predecessor, Collor saw the environment as a diplomatic opportunity, by which the Brazilian government could win international goodwill in exchange for positive acts—goodwill whose effects would spill over into other areas (interview with Fonseca 1990). Nonetheless, even in this area many of the measures taken were simply a smokescreen for a general abdication of environmental governance at the federal level under Collor.

Most of Collor's institutions and policies—again with the exception of the

Table 1.1 Federal Environmental Institutions in Brazil

YEAR	LAW OR DECREE	INSTITUTION CREATED
1967	Decree 6.938	Creates the Brazilian Institute for Forest Development (IBDF).
1973	Decree 73.030	Creates the Special Secretariat of the Environment (SEMA), for pollution abatement and natural resource management, within the Ministry of the Interior.
1981	Law 6.938/81	Creates the National System of the Environment, integrating and reorganizing existing environmental institutions. The National Council on the Environment (CONAMA) is the primary deliberative agency, whose resolutions have the force of law. SEMA is its technical staff.
1985	Decree 91.145	Creates a Ministry for Urban Development and Environment and transfers CONAMA and SEMA to it, continuing their previous functions.
1989	Law 7.735 (February)	Eliminates SEMA and other smaller environmental agencies such as the IBDF; replaces them with a new environmental agency, the Brazilian Institute for the Environment and Renewable Resources (IBAMA).
1989	Law 7.804 (July)	Reorganizes existing environmental institutions (amends Law 6.938/81), and adds a new Superior Council of the Environment as the top environmental decision maker, above the CONAMA, with IBAMA just below it.
1990	Decree 99.244	Reorganizes existing environment institutions, creating a Secretary of the Environment (SEMAM) with four major departments. SEMAM is an autonomous agency below ministerial level, whose secretary reports directly to the president. IBAMA continues as an implementation arm of SEMAM.
1992	M.P. 309 / Law 8.490	Transforms the office of the secretary of the environment into a Ministry of the Environment, and reorganizes its competencies. IBAMA continues.
1993	Law 8.746	Reorganizes the Ministry of the Environment as the Ministry of the Environment and Legal Amazônia. IBAMA continues.

(*Table 1.1 continued*)

YEAR	LAW OR DECREE	INSTITUTION CREATED
1995	M.P. 813	Reorganizes the Ministry of the Environment and Legal Amazônia as the Ministry of the Environment, Water Resources, and Legal Amazônia. IBAMA continues.
1999	various	Removes the primary responsibility for water resources and renames the ministry as simply the Ministry of the Environment. IBAMA continues but is geographically reorganized.

M.P. = Medida Provisória (Provisional Measure), an executive decree law.

Sources: Guimarães 1995; Fernandes 1995; http://www.mma.gov.br/?id_estrutura=8&id_conteudo=928#98, visited 24 July 2005; http://www.seade.gov.br/cgi-bin/titabp.ksh?sg=MAM, visited 9 October 1998.

Amazon—ended after he was impeached on corruption charges in 1992. His successor, Vice President Itamar Franco, moved the environmental agencies into yet another new ministry, the Ministry of the Environment and Science and Technology (Ministério do Meio Ambiente e Ciência e Tecnologia). CONAMA was restored to its previous function as the primary standard-setting body. Franco's successor, Fernando Henrique Cardoso, again reshuffled the environmental agencies when he took office in 1995, creating a Ministry of the Environment and Water Resources (Ministério do Meio Ambiente e Recursos Hídricos), and again keeping CONAMA in a leading role. While some functions have moved in and out of the ministry since, leaving a simple Ministry of the Environment (MMA, with water resources remaining as a secretariat), this formulation proved comparatively long-lasting and continues to this day. This constant reshuffling tells an important part of the story of Brazilian national environmental administration since the transition to civilian rule. The content of the reforms shows a profound uncertainty about the definition of environmental issues: Are they basically urban? About natural resource development—or their conservation? Instead of projecting a holistic vision of environmental concerns, each reform emphasized one piece or another of the environmental agenda. Major institutional restructuring every two years, on average, undermined the agencies' sense of mission and direction. Lines of responsibility and accountability were hard to draw. Once the repeated administrative reorganizations finally stopped, the national environmental agencies could be much more effective in their work and interacted much more responsively and predictably with other actors.

National Environmental Spending

As the environmental bureaucracy expanded, so did spending on the environment, rising from 0.3–0.4 percent of the total federal budget in the 1990s to 0.9 percent in 2005 (see table 1.2). To offer several comparisons, the U.S. government spent 1.3–1.5 percent of the total federal budget on natural resources and the environment from 1993 to 2005. In absolute terms spending in the United States was much higher, rising from $20 to 30 billion, while Brazilian spending was never above $1 billion (Office of Management and Budget 2006, 51–52). Argentina's federal and provincial governments devoted 0.4–0.8 percent of their total public spending to the environment between 1994 and 2000, with total national spending peaking in 1998 at $72 million (CEPAL / ECLAC 2002, 68, 70). Mexico's national spending on the environment topped $1 billion in 2000, 1.06 percent of federal spending for that year (CEPAL / ECLAC 2002, 81).[10]

Interpreting these figures requires considerable caution, especially in Brazil, as they both understate and overstate the federal government's commitment to environmental spending, and say nothing about spending at the state level. As table 1.2 shows, the environment's share of the original federal budget is always considerably higher than what is actually spent. This is because debt repayment and economic stabilization became major government concerns after 1993, and successive administrations used their spending authority to embargo authorized allocations until after they were sure that macroeconomic targets had been met. In practice, this meant that personnel were paid on time and as budgeted, but spending for equipment and programs was always less than the amount budgeted, and federal agencies often received their allotments only in the final months of the year—for a quick frenzy of spending followed by months of budget starvation. In 2003, the first year of the Lula administration, for example, environmental personnel lines received 99 percent of their budgeted amounts and the total federal budget was spent at 91 percent of its original allocation, but nonpersonnel

Table 1.2 Federal Environmental Spending, 1993-2000, 2003-2005

	BUDGETED IN 000 REALS	% OF FED. BUDGET	SPENT* IN 000 REALS	SPENT* IN 000 DOLLARS	% OF FED. SPENDING	FUNDS SPENT AS % OF FUNDS BUDGETED
2005	2,716,319	1.1	2,072,858	888,114	0.9	93.3**
2004	1,562,534	0.7	1,193,443	449,508	0.6	92.2**
2003	2,409,727	1.3	947,749	327,374	0.6	91.1**
2000	1,953,382	0.7	1,274,004	653,039	0.5	65.2
1999	853,434	0.7	726,451	401,576	0.3	85.1
1998	1,226,239	0.4	929,570	769,193	0.4	75.8
1997	1,089,172	0.4	836,173	748,927	0.4	76.8
1996	1,132,036	0.5	819,461	685,597	N/A	72.4
1995	1,941,767	0.7	666,606	N/A	N/A	34.3
1994	1,755,734	0.4	N/A	N/A	N/A	N/A
1993	1,660,335	0.5	N/A	N/A	N/A	N/A

* This is *despesa liquidada*, meaning that a good or service has actually been ordered and delivered.
** Figures represent percentage spent of the entire federal budget for the year.

Sources: Figures for 1993-2000 based on Young and Roncisvalle (2002), tables 2.4, 2.5, 2.6. Figures for 1993-2000 are in 2001 reals; those for 2003-5 not adjusted for inflation. Since amortization of debt was included in ministerial budgets for 1993-2000 but not later ones, amortization was subtracted from the totals reported by Young and Roncisvalle. All conversions to U.S. dollars based on exchange rates as of 31 December of the relevant year. Figures for 2003-5 calculated from data available on consulta.tesouro.fazenda.gov.br/cofin/dotacao_vs_desp_param.asp.

environmental spending was limited to just 30 percent of its original allocation.[11] During the currency crisis of 1998-99 President Cardoso also cut programmatic environmental spending first, and disproportionately (Kasa and Næss 2005). These developments suggest that the federal government's commitment to environmental spending is less than the apparent secular rise would indicate.

However, a focus on the national budget underestimates the resources available for environmental programs at the federal level in Brazil. The national treasury contributed only 32.3 percent of the 2003 budget for the Ministry of the Environment, and that share was largely tied to obligatory personnel spending. International credit and donations accounted for an additional 13.3 percent of the ministry's budget. The remainder came from domestic sources, especially water-use concessions, compensation for water, oil, and natural gas, and environmental fines (Souza 2002, 4–5). Most environmental projects and investments are actually funded by non-treasury sources. These help make up for the shortfalls noted above. During the currency crisis of 1999, for example, a coalition of domestic and international NGOs and public actors successfully pressured the government to continue to accept new international environmental loans, and as a result could help to sustain additional funding and programs (Kasa and Næss 2005). This sort of action has added considerably to the resources available for environmental protection in Brazil, though at some cost to the Ministry of the Environment's ability to set priorities and organize its own programs (Barretto Filho 2003; Souza 2002). The ad hoc character of program funding is an important element of continuity with the much smaller SEMA of the 1970s. As we noted in the Introduction, this fluidity confounds efforts to make clear analytical distinctions between state and nonstate, domestic and international, in Brazilian environmental politics.

Environmental Agencies and Citizen Participation

Besides budgetary politics, nonstate actors have found their way into the process of environmental protection in a variety of areas. More systematic opportunities for citizens' groups and social movements—mostly domestic —to participate in the process and influence environmental policy have varied with shifts in institutional framework and personnel at the top. With democratization, a politicized environmental movement wanted to influence the ends, as well as the means, of environmental policy. SEMA's technocratic stance, which may have gleaned for it a measure of autonomy and authority during the military period, became counterproductive with the return to civilian rule, discouraging the participation of environmentalists who may have lacked technical training but possessed considerable situational knowledge.

During the Sarney administration (1986–89), the main avenues of participation were through CONAMA and (beginning in 1989) the National Environmental Fund. However, environmentalists complained that they were marginalized from both. Environmental representatives were sometimes invited to meetings late and found that they were expected to ratify decisions already made, with little information or input on their part (interviews with Aveline 1990; Iglesias 1991; Petrillo 1990; Schäfer 1990). Clearly, formal representation was not sufficient to produce real influence. Nonetheless, environmentalists were outraged when Collor's reforms in 1990 eliminated those participatory opportunities, however imperfect. Dozens of environmental groups from all over the country jointly remonstrated against CONAMA's demise and the rise of the new Council of Government and its strong state control, but they were unable to gain political traction until after Collor was impeached and his changes generally overturned by his successors.

In 1998 environmentalists from around Brazil gathered at the State Environmental Secretariat in São Paulo to discuss the environmental councils at various levels of government and to evaluate them as spaces for environmental activism. The discussions were sponsored by PROAONG, a department set up within the secretariat to facilitate access by NGOs to state government and coordinate information for and about environmental NGOs in the state.[12] At various points in the discussion, environmentalists labeled their frequent griping about the councils' functioning as a *choradeira* (roughly, pity party). Despite their dissatisfaction, when João Paulo Capobianco suggested that it was time to move on to other strategies, everyone else present stuck to the position that participating in the councils was worthwhile (Svirsky 2002).

Many of the environmentalists' complaints centered on their own inability to fill the democratic space they had successfully demanded. Their experiences suggest that the recent enthusiasm for popular participation in environmental decision-making (see e.g. Durant, Fiorino, and O'Leary eds. 2004, part II) may face significant logistical limits. In Brazil this is in part the result of the proliferation of councils. One organization held seats on more than ten councils, and another simply reported being on "innumerable" councils (Svirsky 2002, 32, 45). With many environmental organizations still relying on volunteers, the council commitments weigh heavily on limited time and resources. Environmentalist participants admitted that they often went to meetings with little time to read supporting materials, no opportunity for consultation with other environmentalists, and even less time for reporting to others what they had done. As a result, they participate primarily as individuals, representing at best a generalized set of beliefs and interests of their fellow environmentalists. Many participants also felt unprepared to evaluate all the materials, and some expressed resentment at being asked to be unpaid technical consultants for a government that provided no independent technical staff for the councils. While the scientific dimension of environmental questions makes such problems especially sharp for environmentalists, many of their comments match those of participants in other kinds of councils, such as those on health, children and adolescents, and social assistance (Tatagiba 2002).

The group of environmentalists in this discussion of environmental councils was drawn from those registered with CONAMA and active at the national level. If they consider themselves unprepared, the vast majority of environmental organizations must feel even more so. In most states only a minority of environmental organizations even registers with the state councils, which would be necessary for them to be selected by their peers as participants. In São Paulo under 20 percent of the NGOs that are registered with PROAONG also register with CONSEMA. Only 169 NGOs were registered with CONAMA in 1991, a high point in environmental mobilization (interview with Svirsky 1999). Typically the same organizations appear over and over in the national, state, and municipal councils. These factors limit environmentalists' ability to walk away from the councils in protest of problems. In both Santa Catarina and Mato Grosso states, governors happily accepted alternative environmental organizations that they could control when historic activists left the councils—and environmental licensing proceeded apace (Svirsky 2002, 34–35, 43–44).

Environmentalists are even more dissatisfied with the national-level CONAMA. For example, they were determined that CONAMA should not be limited to a role of environmental licensing. Miguel Scarcello of Acre said, "We go there with the impression that it is a major forum, where we will be able to deliberate on the decisive questions of the country. Unfortunately, when we get there, we find a routine that constantly ends up taking us out of policy discussions to very topical issues" (Svirsky 2002, 74). Environmentalists bear some of the blame for placing quite local concerns on CONAMA's agenda (Capobianco 1997, 11). In their view many governmental decisions and policies that should go through CONAMA do not.

During the discussions of CONAMA in 1998, environmentalists were especially indignant about the way a revision to the environmental impact assessment law of 1986 had just been pushed through the council. The technical requirements, timetables, and costs of obtaining an environmental permit were all reduced (Campanilli and Leitão 1998). Environmentalists felt that their presence had legitimated a decision they did not support—not an unusual outcome in democratic politics—but an especially bitter loss for them as it would have such broad effects and they were part of the discussions but unable to influence them. Yet they were hesitant to give up their participation: "The Council is still a way out, and with it we at least always have a way of provoking the public power. Whether they like it or not, we come in there, we do and say what we want, and it happens" (Svirsky 2002, 66).

These are not new complaints. In fact, there is a striking similarity between the criticisms raised by participating environmentalists in 1998 (Svirsky 2002) and those raised almost a decade before (interviews with Iglesias 1991; Petrillo 1990; Schäfer 1990). The sense that the national government impedes meaningful deliberation persisted as well, exacerbated by several formal decisions in the mid-1990s to move some areas of debate out of CONAMA's purview, transferring them to other bodies to which NGOs had less access (Capobianco 1997, 10).[13] The environmentalists' negative evaluation is somewhat belied by the literally hundreds of resolutions passed by CONAMA over more than two decades, which combine to give Brazil a comparatively strong regulatory foundation on the environment. Just a few items on CONAMA's agenda in the early twenty-first century included oversight of genetically modified organisms (which CONAMA seems to have lost in the larger policy debate—see Hochstetler forthcoming) and conservation policies on private land.

In any case, licensing conflicts have increasingly moved to a new battlefield, the courts—again, at all levels of government. While businesses had always exercised this option, the Ministério Público is now a frequent "judicializer" of the licensing process, using the courts to insist on environmental impact assessments that have been omitted, or questioning the quality of completed ones (Moraes 2005, 209). These proceedings are all the more weighty, as Moraes notes, since the licensing process remains open to public scrutiny (having been established at the height of democratizing efforts during the transition), and therefore has become virtually the only setting that requires public debate about economic decisions, even state projects. As a result, individual licensing decisions often become crucibles for the airing of giant social conflicts that really should be settled in other ways, but are not (Moraes 2005, 220–21). That environmental activists have had an institutionalized seat at the table for twenty years to discuss such conflicts, and citizens have been able to join these processes as well, mean that environmental decision making is in fact considerably more open to public scrutiny in Brazil than in many countries, whatever the limitations and final decisions.

The Constituent Assembly

Besides the councils, the institutional conditions for environmental protection from the 1980s on were also reshaped by changes in the legal structures, beginning in 1981 with the legislation discussed above, and continuing with two other important changes: the insertion of environmental provisions into the new Constitution promulgated in 1988; and the activation of the Ministério Público and a law in 1985 providing for civil suits in the public interest to attend to environmental concerns.

Environmentalists had an unexpectedly significant impact on the National Constituent Assembly, made up of the combined Chamber of Deputies and Senate elected in 1986, which wrote Brazil's eighth Constitution in 1987–88. The only environmentalist candidate elected to the new Congress, Fábio Feldmann (PMDB–São Paulo, later PSDB), created opportunities for environmentalists to make their case in Brasília, by holding a series of hearings on how the environment should figure in the new Constitution.[14] By contrast, the Subcommittee on Health, Social Security, and the Environment devoted only one of its official hearings in May 1987 to the environment (*Estado de São Paulo* 14 May 1987). The subcommittee heard testimony from OIKOS, an environmental group in São Paulo associated with Feldmann, the Brazilian Society for the Advancement of Science (SBPC), SEMA and CONAMA, and the Institute for Studies of Contemporary Community Problems.[15] Feldmann organized a series of congressional visits to six states between April 1987 and March 1988, to acquaint members with a broad range of national environmental concerns. During these visits, environmentalists showed up at public hearings to offer advice; at a hearing in Ceará in

October environmentalists made ninety-four suggestions for inclusion in the new constitution (Frente Nacional 1988, chapter 6).

The goal of these hearings and debates was to develop materials to justify a constitutional chapter on the environment. Feldmann's preliminary draft addressed key environmentalist demands, some of which were proposed directly in a popular initiative process (Micheles et al. 1989). This draft prohibited virtually all nuclear production; gave joint responsibility for the environment to municipal, state, and federal governments; required environmental impact studies, with public comment and evaluation; guaranteed citizens' access to information about the environment; labeled environmental infractions as crimes and genocide; declared major ecosystems to belong to the national environmental patrimony; provided mechanisms for citizens to take free legal action against environmental infractions; and guaranteed the right to a safe and healthy environment (Frente Nacional 1988, chapter 19).

Despite its breadth, the draft chapter came through its first review nearly intact. Of the 104 amendments that the Subcommittee on Health, Social Security, and the Environment considered, over half related to the articles banning nuclear weapons and nuclear energy production, allowing nuclear reactors only for scientific ends (Câmara dos Deputados 1988). Gastone Righi (PTB–São Paulo), who was identified with the multinational corporation Rhodia, also led a spirited fight against the environmental crime provision (*Jornal da Tarde*, 26 May 1987). However, in the final draft approved on 25 May, the subcommittee had added only four articles to the original eight, and rejected only the draft's prohibition of nuclear production for peaceful ends.

More opposition appeared at the next stage, from powerful lobbies like the Brazilian Chemical Industry Association, supporters of nuclear energy, and producers and users of pesticides (Frente Nacional 1988, anexo III). With funding from SEMA, Feldmann convened a meeting of environmentalists in Brasília on 5 June 1987, hoping to mobilize them for further action. Representatives of seventy-one organizations formed the National Front for Ecological Action in the Constituent Assembly (National Front), to offset the anti-environmental lobby. A highly diverse group, it included the activist state environmental federations (APEDEMAS) from São Paulo and Rio de

Janeiro, the Brazilian Society for the Advancement of Science (SBPC), prominent individuals like Nogueira Neto and Maria Teresa Jorge Padua, and several Green Party activists. They were enthusiastic that Feldmann had included their most controversial proposals in his draft, and they were ready to help him push the proposals further.

Feldmann invited prominent legislators to address the meeting, including Ulysses Guimarães, president of the Constituent Assembly, and Bernardo Cabral, whose Integration Committee (Comitê de Sistematização) would put together the chapters generated by specialized committees into a single constitutional document. Attention from such prominent legislators seemed an encouraging signal that the environmental chapter would be taken seriously. Environmentalists eventually signed up nine senators and eighty-two federal deputies as members of a nonpartisan, multiparty "Green Front" (*Estado de São Paulo*, 9 June 1987), albeit comprising many shades of green (*Folha da Tarde* 9 May 1988).

The first hurdle was discussion in the full Committee on Social Order. Feldmann had timed the creation of the National Front to demonstrate support just before the committee vote on the chapter. Opponents proposed a new round of debilitating amendments pertaining to the nuclear question, the number and status of conservation areas, and environmental impact studies. Nevertheless, on 15 June the committee approved a chapter on the environment that retained the concept of ecological crime and looked quite a bit like the subcommittee's version.

Next, the Integration Committee produced three drafts of the entire Constitution, in July, August, and September. At each stage plenary discussions considered literally tens of thousands of amendments. Most of the chapter's provisions were retained during this phase, although some were moved to other parts of the Constitution. For example, the right to file a Popular Action suit without cost because of environmental damage was moved to the section on individual and collective rights. Those writing the drafts wielded great influence, because the language they used often received minimal attention for logistical reasons. The chapter on the environment simply passed through several stages in its draft wording—which followed the committee wording closely—without floor debate.

Meanwhile, environmentalists continued their pressure by both lobbying and protest. In November ten thousand people marched against the nuclear installation in Aramar, São Paulo. In December several hundred environmentalists, scientists, members of Congress, and public employees from the national environmental bureaucracy mobilized in Brasília to support a whaling ban. This gathering showed that environmentalists had a broad repertoire of action on which to draw: on a single day, the same groups of people testified as environmental experts on the content of the environmental chapter, lobbied legislators on constitutional and policy issues, and finally held a rally of three hundred people at the National Congress while carrying a giant inflatable whale, in violation of a resolution forbidding demonstrations at the Congress. Environmentalists scuffled with guards, and Feldmann was threatened with a lawsuit over criticisms he made of the president of the Senate. Legislation to ban whaling was passed later that month in limited form, but constitutional success took longer.

As the constitutional discussions moved into a new round of amendments at the end of January 1988, a new center-right coalition called the Centrão, or Big Center, appeared to threaten many of the gains of both environmentalists and the Pro-Participation Plenary, an organization formed to ensure that any new constitution would provide for maximum participation by citizens (Michele et al. 1989). The Centrão's main objectives, aside from supporting Sarney's bid to lengthen the presidential term, were to weaken the constitution's labor rights provisions and reduce restrictions on businesses and property owners (Martínez-Lara 1996; Micheles et al. 1989). Since amendments with 280 congressional signatures (which the Centrão could muster) got priority in the voting order, the coalition could control the Assembly's agenda. It also produced an entire alternative text of the Constitution, and proposed many amendments by simply demanding substitutions of its text for the original.

The first sentences of the Centrão's environmental chapter neatly summed up the difference between the two versions (Frente Nacional 1988, anexo II). The Integration Committee drafts began, "Everyone has a right to an environment in ecological balance, it being the responsibility of Public Authority and the collectivity to preserve and defend it." In contrast, the Centrão version began, "The Public Authority will protect environmental and ecological balance, as a means to ensure the quality of life and the protection of nature." By eliminating the collectivity's responsibility for defending the environment, the National Front understood the Centrão to be implicitly denying its *right* to do so. The Centrão draft also weakened the chapter's substance, for example by retaining the call for environmental impact studies, but not the requirement that the results be publicly available. It abandoned the language of conservation in favor of the language of "rational use." The Centrão draft also eliminated provisions for popular initiatives or referenda that were present in other parts of the Integration Committee draft (*Folha de São Paulo*, 13 January 1988). After having lost votes on the nuclear issue and on public interest limits to the exercise of property rights, environmentalists returned to Brasília on 5 May 1988 to march and lobby for the Integration Committee's version over the Centrão's.

Their first big victory came on 25 May, when the Assembly designated important natural areas—the Atlantic Forest, the Amazon, and others—as areas for conservation rather than rational use (*Noticias de Poá*, 4 June 1988). Most other pieces of the environmentalists' proposal eventually passed as well, including penal and administrative sanctions for environmental degradation as well as responsibility for reclamation of any damage. On matters related to citizens' participation, language that environmentalists supported also defeated that supported by the Centrão. The chapter on the environment guarantees Brazilians a healthy and sound environment and commits the public as well as the government to this end. Mechanisms of participation were preserved in the environmental chapter and in the Constitution as a whole (Moisés 1990).

The final draft of the chapter on the environment passed the first floor vote, with 450 Assembly members voting in favor, three against, and four abstaining. This was the highest positive vote total of any piece of the Constitution to that point. The chapter passed the second vote with similar levels of support. Success came in part because the Centrão itself was split over environmental questions and at least four of its members participated in the National Front as well, notably President Sarney's son José Sarney Filho (*Folha da Tarde* 9 May 1988). A decade later he would provide solid leadership as minister of the environment and eventually join the Green Party. On 22 September, almost twenty months after the Constituent Assembly began, Brazil had its eighth Constitution, its first ever with a chapter on the environment.

The constitutional language on the environment is notable less for introducing new ideas than for providing a strong underpinning to existing legislation and preventing efforts to roll it back. Even before the constitutional process, Brazil had "a wide range of flexible and interesting legal instruments, which were sufficient to give legal support to public policies and private actions on the matter of environmental protection" (Fernandes 1995, 92–93). The environmental chapter withstood the mandatory review process in 1993 (*Viva Alternativa* 25 (1993)), and to date no significant changes have been proposed; it has proved a solid foundation for many environmental rights and protections. In the 1990s numerous pieces of legislation built on its premises, providing Brazil with unusually strong foundations for environmental law compared to its South American neighbors (Hochstetler 2003).

The (Un)Rule of Law

While Brazilian environmental law is ample and often well formulated, those characteristics are not enough to guarantee its effective application. When Brazilians are asked about particular laws and their impact, one possible response is, "That law never caught on" (*Essa lei não pegou*). The phrase captures the frequent gap between legal and substantive reality. The circumstances under which a law takes hold (*a lei pega*) have not been well studied. We might imagine that a law would need to have sufficient political support, encourage voluntary compliance, and establish clear sanctions for noncompliance (whether the law pertains to public authorities or private parties). It should be easy to monitor the behavior regulated. And it must be possible to prosecute and sanction violators of the law, whoever those might be. None of these are easy conditions; they are especially difficult in the Brazilian context, where the rule of law has always been tenuous and its application profoundly unequal. As a result, public authority owes considerably more to the exercise of power than to the binding nature of the law (O'Donnell 2001; O'Donnell 2004). The grotesque levels of social inequality that affect other aspects of Brazilian social and political life are strongly reflected in the legal system, summarized in an old Brazilian saying attributed to Getúlio Vargas—"For my friends, anything—for everyone else, the law."

This situation of course has a noxious effect on Brazilian politics generally, and is made even more problematic by the growing involvement of gangs in drug trafficking and other illegal activities in many parts of the country. Although this state of affairs does concern us (most directly when we discuss the Amazon in chapter 4), here we make only one observation: the passage of an environmental law or issuance of a regulation is but the opening move of the political struggle required to make it meaningful. The implementation struggle is at least as important as the effort made to get it passed in the first place. This is true of both ordinary and constitutional law.

The deficiencies and sometimes outright corruption of the Brazilian judiciary (understood as the courts, the legal profession, and the various police forces) have been lamented for decades, and have proved remarkably resistant to reform (Holston and Caldeira 1997, Kant de Lima 1995; Pereira 2000; Pereira 2006). The problems associated with the courts are both social and institutional. Sadek in her magisterial review of the problem of judicial reform in Brazil argued that the courts have been used intensively by those who knew how to benefit from their use (Sadek 2004, 24–25). Part of the difficulty in achieving meaningful reform has been that these people derive distinct advantages from a judicial system that makes the courts virtually inaccessible to most people because of its byzantine procedural requirements, is characterized by proceedings that drag on interminably, fails to establish binding precedents, and allows for almost unlimited appeals on almost any grounds. The enforcement of legislation that thwarts powerful interests, as environmental legislation may frequently do, is not at all guaranteed. In December 2004 the Congress passed Constitutional Amendment 45, which affected both the structure and the functioning of the judiciary, making possible a variety of pending legislation designed to speed up the judicial process—for example, the requirement that decisions of the Supreme Court have binding effect on identical cases (Law 11.276, 7 February 2006). Although it will take some time before the effect of judicial reform is felt (at the end of March 2006, the Supreme Court had 199,503 cases on its docket), some elements are already being put in place (OAB 2006).

Even before progress had been made on judicial reform more generally, during the 1980s two interrelated developments in the Brazilian judiciary promised to open up legal avenues for greater vigilance over the enforcement of laws and the protection of collective rights. One was the reorientation of the Ministério Público and its public prosecutors away from criminal prosecution and increasingly toward civil suits; the other was the passage in 1985 of a Law of Public Interest Suits (Lei de Ação Civil Pública, Law 7437), which gave the Ministério Público far greater latitude to intervene in defense of a diffuse public interest. The Ministério Público had always been present in civic lawsuits in situations characterized by the unavailability of rights or the incapacity of rights-bearers, but the Civil Process Code of 1973 (Código do Processo Civil) charged the institution with defending the public interest more generally (Art. 82, para. III) (Arantes 2002, 25–31). In 1981 Lei Complementar 40, known as the Organic Law of the Ministério Público, began the process (continued with the 1988 Constitution) of freeing the institution from its previous subservience to the executive branch, making it the monitor of administrative acts and the protector of the public interest. The SISNAMA of 1981 empowered the Ministério Público to act for what are referred to as "meta-individual" interests, among which are diffuse interests, collective interests, and "homogeneous" or group interests (on the distinctions among these see McAllister 2004, 91 n. 139). This expansion of collective rights offered a fertile new field for action in the public interest, and by mid-decade *promotores públicos* (somewhat akin to district attorneys) had begun to use them.

During the 1980s jurists in São Paulo especially were engaged in a lively debate over the drafting of a law enabling and regulating the terms of public interest lawsuits. The debates over this law reflected contending ideas during the democratization period regarding individual and collective rights, the role of intermediary associations, and the need for a tutelary role for the state (in the form of the Ministério Público) as a guardian of the interests of those who could not speak for themselves (Arantes 2002). The resulting Law of Public Interest Suits assigned a much more central role to the Ministério Público than many of the jurists in São Paulo had wanted to see, continuing the guardianship role. Although the law also gave standing to sue to legally constituted organizations whose charters or statutes contained language justifying their claim to speak for the diffuse or collective interest being harmed, exercising it proved quite difficult in practice.

Law 7437 gave the Ministério Público the power to launch at its own initiative a civil inquiry, in which parties (public or private) were required to provide the information requested. This made it a much stronger—and more politically powerful—actor in the system. Because prosecutors would thus have an easier time investigating cases of abuse than any civic association could have, it made much more sense for citizens' groups to refer problems to the Ministério Público from the beginning, instead of attempting to investigate the problems themselves. In a study of prosecutors in the state of São Paulo, the anthropologist Cátia Silva found that especially entrepreneurial prosecutors intervened in the community with an active transformational agenda, seeking out problems and using their powers in the Ministério Público to resolve conflicts (normally outside the courts system, whose response is understood to be both slow and uncertain; Silva 2001, 137–38). A decision by prosecutors to investigate a case to determine whether a public civil suit is merited is often itself sufficient to persuade the parties to negotiate an agreement. Such legal agreements, called Conduct Adjustment Agreements (*compromisos de ajustamento de conduta*), are potentially strong instruments, with measurable steps to be taken and legal consequences for their violation.

To the prosecutors in the Ministério Público, the law's implicit maintenance of the tutelary relationship between state and society (Arantes 2002, 76) made perfect sense. Environmental organizations, for the most part unschooled in the law and unaccustomed to moving through the procedural minefields of the Brazilian legal system, were simply not equipped to undertake these tasks. Asked about use of the public civil suit by environmental organizations, the prosecutor Edis Milaré responded: "Environmental organizations have participated in fewer than 5 percent of them. Maybe it's even a matter of convenience. Convenience in the sense that generally [the organizations] are not very well structured. They don't have a legal department at their disposal. So they work more on consciousness raising, but when they need to resolve a problem they send it to the Ministério Público, which, for better or worse, is reasonably well set up to do this" (interview with Milaré 1991). Overall the Ministério Público has played an even more central role in environmental public civil suits, of which it has filed 97 percent of the total, than in other areas (McAllister 2004, 229). It is fully "the environmental lawyer and negotiator of society" (230).

One notable exception, an early case in which an organization did play an important role, was the lawsuit initiated in 1986 by the environmental organization OIKOS against the companies charged with polluting the city of Cubatão. Feldmann, not yet a congressman, had been among those lawyers who tried to use public civil action provisions predating 1985 for environmental purposes, bringing suit through the Ministério Público on behalf of the environmental organizations São Paulo Association for Nature Protection (APPN) and later OIKOS (see chapters 3 and 5) against the directors of a polluting firm in São Paulo called Aliperti. They lost, having accused the firm of emitting dangerous gases, a crime under the 1939 penal code, which provided no penalties for environmental crimes. Feldmann applauded the diffuse interests law of 1985 both for creating a new type of lawsuit and for opening up the possibility of legal standing for civil society. The case against the industries of Cubatão was a watershed because of its size; interviewed in 1991, Feldmann believed that it would show the importance of the new law as an instrument of civil society (interview with Feldmann 1991).

However, almost twenty years later the case remains in the courts, a remarkable example of the ability of those with the resources to continue appealing to enforce nondecisions. On average, 54 percent of parties to decided public civil suits waited more than two years for their first judgment. However, the decisions have often been worth the wait: the first judge accepted the environmental argument entirely in half the cases, and partially in another 17.5 percent of cases. Further, 80 percent of the cases were upheld on appeal. Judges also frequently granted injunctions (*liminares*) when requested—57 percent of such requests were fully granted and another 6 percent were partially granted (Instituto Socioambiental 2005).

Although the new law did not become an instrument for a more autonomous civil society, it did motivate environmental groups (and eventually consumer groups, women's groups, racial minorities, and others) to collaborate with the Ministério Público. An interesting feature of the Ministério Público is that its investigations follow the principle of obligatory action: a prosecutor who becomes aware of a possible harm must investigate the complaint and pursue it to restitution or to the conclusion that no harm has in fact been done (McAllister 2004, 132). In São Paulo's state-level Ministério Público alone, prosecutors carried out 21,000 investigations between 1985 and 2001 (McAllister 2004, 136), with 77.5 percent of the investigations between 1990 and 2001 based on complaints from third parties (232).

Along with its new attributions, the Ministério Público also began to create specialized divisions for different kinds of problems. By 2002 there were more than two thousand prosecutors who specialized in environmental matters across Brazil, with about two hundred working on them almost exclusively (McAllister 2004, 123–24). In state capitals there were usually specialized staff members who could assist other prosecutors with a growing number of environmental cases. From 1984 to 2001 prosecutors in São Paulo's state Ministério Público alone filed an average of 176 environmental public interest suits annually. Even this number was dwarfed by the much larger number of Conduct Adjustment Agreements, which reached about a thousand in 2000 (151–52). The passage in 1998 of an Environmental Crimes Law increased the penalties for environmental damage, but the requirement of proof of intent to cause damage has made it difficult to use, causing the Ministério Público to continue to rely on civil suits and nonjudicial remedies. Prosecutors have come to see both the Environmental Crimes Law and the public civil action suits as most useful for preventive purposes and as the backdrop for effective negotiations with degraders of the environment (175, 242).

The development of these legal tools has been one of the most important —and distinctive—components of Brazilian environmental politics over the last two decades, one crucial to many blocking activities. Only the United States has made such extensive use of legal strategies for environmental protection. Activists have used them to block or slow down many potentially harmful proposals, from a water superhighway down the La Plata River to the cultivation and sale of genetically modified food (Hochstetler 2002a; Hochstetler forthcoming; McAllister 2005). They have made both environmental legislation and environmental agencies more effective than those who crafted them may have anticipated, despite ongoing limitations. The tools remain adversarial, however. As a virtual fourth branch of government, the Ministério Público uses its ability to demand information and question the activities of governmental as well as private actors to function as an agent of horizontal accountability, with all the benefits that this brings in a country where accountability has come quite late to government agencies (Campos 1990). At the same time, even environmental agencies often chafe under what they see as unnecessarily legalistic interpretations that limit their ability to adapt to circumstances, and have found the workload of responding to those thousands of investigative requests a real burden that takes time from other activities (McAllister 2004, 211–15).

Trajectories

Throughout the institutional processes we have discussed here, there has been an underlying story of particular individuals and groups attempting to change both the rules and the behavior of formal institutions, and to make use of whatever channels were available to affect processes in course. With democratization came what appeared to be a confrontation between on the one hand those calling for radical changes in political institutions and social structures and on the other hand those who sought formulas by which a regime transition could be effected without alienating either the military (which might decide not to leave power after all) and those who had supported the authoritarian regime. Although many writing about the Brazilian transition to democracy interpreted these tendencies in terms of the political choices available during the transition itself, these are recurring themes in Brazilian history, what Antônio Candido in his preface to Sérgio Buarque de Holanda's *Raízes do Brasil* categorized as rigidity versus conciliation (xxii). Indeed, with each new "critical juncture" of Brazilian history, the old has been reincorporated into the new, and differences have been eliminated by treating them as if they did not exist.

The skills necessary to navigate in such unmappable territory are learned through experience; they are accumulated by individuals rather than institutions, although they can be learned collectively; and individuals carry them from one institutional setting to another (Mische 2006). Although we are not referring only to formal settings, it is in these settings that such skills are especially useful, since policy making and implementation require collaboration among a variety of organs, and changes in rules, norms, and behavior cannot be enforced coercively. The importance of these accumulated skills is evident in the story we have told about the development of federal environmental capacity.

Appointed under the military regime, Paulo Nogueira Neto used personal connections with some of Brazil's most powerful media elites to promote the environmental cause. Those connections were available to him because of his family's political trajectory as anti-Vargas political exiles. This history created a series of assumptions about his connections and allegiances that we can speculate might have disposed military leaders to accept him when his name was suggested as environmental secretary; the anti-Vargas UDN party had provided the military with some of its most stalwart early supporters. Beyond his personal links and his professional ties in Brazil as a professor at the University of São Paulo and a longtime conservationist, he also had a long association with international conservation circles. That he became a member of the IUCN executive board in 1970, well before Stockholm, demonstrates his credibility in those circles. International credibility in turn helped with mustering small quantities of foreign money for the secretariat's projects. Nogueira Neto was an early example of someone that Sidney Tarrow (2005) would call a "rooted cosmopolitan."

There is another key point in understanding how individuals' trajectories matter in the development of institutions: relationships developed in the course of one set of activities build trust that carries over into new ones. Here the links between Nogueira Neto in SEMA and Maria Teresa Jorge Padua in the IBDF—the institution on whose territory he encroached most extensively—prevented a rivalry from becoming a destructive feud. Padua had long fought within the IBDF for more serious efforts to conserve forests. Like Nogueira Neto she was active in the IUCN, elected in 1976 to be the Latin American representative on its Commission on National Parks and Protected Areas. She first met Thomas Lovejoy at the IUCN's 12th General Assembly in Zaire, where tropical rainforests headed the agenda, and the two collaborated on projects linking scientists from the United States and Brazil. In 1980 they wrote an article on refugia for the *IUCN Bulletin* (November–December, 11–12). In the early 1980s Padua and Nogueira Neto had fought together against the construction of a road in the Araguaia National Park and in 1982 they jointly won the Getty Prize, awarded by the World Wildlife Fund. But by 1982 Padua had given up believing that the federal government wanted to control the devastation in national parks. She and her team presented the president of the IBDF with a document showing that almost all the twenty-four national parks were being destroyed: IBDF stamps and signatures were being used by smugglers of endangered animal skins; more than six hundred forest guards did not earn enough to buy uniforms; boats and jeeps rarely had fuel, and there was no ammunition for the guards' revolvers; some were killed by smugglers and others had become complicit with them (*Jornal da Tarde*, 13 December 1982).

Similarly, Fábio Feldmann carried into the Chamber of Deputies the web of relationships and credibility he had developed through a decade of participation in the environmental movement in São Paulo. As a young prosecutor trying to hold firms legally responsible for environmental damage, he participated in debates on the Law of Public Interest Lawsuits. By the mid-1980s he was one of those convinced that the environmental movement's only hope was to professionalize, seeking resources for civil society organizations to act on their own. With his congressional mandate as part of the picture after 1986, Feldmann and his group were able to think strategically about how, for example, certain causes were better advocated by oikos, others by the sbpc, still others by his congressional office, and so forth. This close-knit network shaped Feldmann's actions in Congress, insofar as he put his office at the service of the environmentalist cause rather than his own party or career (interview with Feldmann 1991). In other words, Feldmann's actions should be understood in terms of a *project*, a projection from present to future in which acts are the execution of a projected act, part of a process involving reflection (Schultz 1967, 57–63; Mische 2006 forthcoming). Feldmann was strategically positioned at the junction of a variety of institutional fields—the arena of environmental movement activism, a project-based network, an environmental organization (oikos), political partisanship (the pmdb and psdb), the legal community and bar association, São Paulo's Jewish community, and the Federal Congress. He was positioned to act as a mediator among these fields, although his personal trajectory and project focused his mediation on the interaction among movement, projective, legal, and congressional fields. In his later post as secretary of the environment in São Paulo, Feldmann found that the fit between his environmentalist trajectory and the state technical bureaucracy was more problematic, serving his project well in some instances but undermining it in others, as we will see in chapter 5.

José Lutzenberger could have been expected to bring similar kinds of skills and projects to Brasília when he became Collor's surprise choice for environmental secretary. He was, after all, a pioneer of the "new environmentalism" in Brazil: his leadership in struggles against pollution in

Rio Grande do Sul, his writings and lectures inspiring people to take action in other parts of the country, and his close relations with environmental and technical communities in Europe (especially Germany) and the United States made him larger than life. Yet Lutzenberger was not a mediator—he was a visionary. He had never learned, nor been interested in learning, the skills that would have allowed him to move easily among diverse fields of action, while emphasizing or downplaying various aspects of his identity to enhance the conversation (Mische 2006 forthcoming). His role had been inspirational and technical, much more than organizational or strategic.[16]

Uninterested in the day-to-day functioning of his secretariat, Lutzenberger was an indifferent administrator at best, serving primarily (as Collor and his advisors had hoped) as a foreign relations asset, a powerful signal to foreign observers that the government was serious about attacking environmental abuses. As he became increasingly out of touch with the Brazilian context, the secretary moved his headquarters to a national park, and was much more available to foreign environmentalists and journalists than to Brazilian ones. Only Amazon activists had better access to the secretariat than they had ever had before. Lutzenberger's close ties with Mary Allegretti, then head of the Amazonian Studies Institute in Paraná and collaborator of Chico Mendes in Acre, and with others involved in the Amazon, represented a partial exception to his failure to involve Brazilian environmentalists. Although his undistinguished tenure undoubtedly resulted partly from budgetary restrictions and a lack of genuine support from higher levels of government, these factors were exacerbated by Lutzenberger's lack of mediation skills. His failure to mediate effectively (or at all) among the myriad fields to which he had access made his passage through the federal bureaucracy a missed opportunity.

Although the people discussed above exemplify the important relations between the trajectories of individuals and the functioning of institutions during the construction of federal environmental organs in Brazil, there are many others who have long played mediating roles in lesser capacities. We will meet some of them in later chapters. Brazil is certainly not alone in owing a great deal to the leadership, entrepreneurship, and custodianship of dedicated activists and administrators; however, the peculiarities of Brazil's simultaneously rigid and extremely bendable institutional structures make it especially important for us to recognize how these contingent forms of individual trajectories, projects, and skills combine to produce structures—rules and roles—that will intersect with the trajectories of those who follow them.

Since the Brazilian military government created a tiny, three-person national secretariat to deal with environmental issues in 1973, Brazil's formal environmental capacity has been radically transformed. The national environmental agencies themselves were reconfigured many times, gradually gaining personnel and responsibilities until they took the form of the full-fledged institutional structure that we see today. Environmental legislation was written throughout the period, by the military regime and then by its civilian successor governments. The 1988 Constitution devoted a chapter to the environment; other innovations not specifically about the environment gave Brazil a powerful "fourth branch" of government in the form of the Ministério Público, empowered to defend collective rights like the right to a healthy environment in the courts and through extra-judicial agreements—even against the state. All these changes were the result of political struggles by a variety of primarily domestic actors, both state and nonstate (and sometimes both at once). In an unusually protracted transition from military to civilian government, they managed to keep the environment on the agenda and moving forward. The skills and contacts of particular actors were central in the unfolding of this story, especially in the early stages of Brazilian environmental politics.

Underneath this rosy tale of remarkable changes, however, environmental proponents confronted the undertow of Brazilian politics, with the powerful often above the law and the drive to develop trumping most other aims. Informal and personal arrangements undercut formal institutions and formal equality. Budget constraints limited the resources available for environmental conservation even as agencies institutionalized. These more permanent features of Brazilian political culture survived the end of military rule. The undeniable pluralization of voices during the transition—to include subnational governments, environmental activists, ordinary citizens—did not guarantee that their clamor would be heard, and the multitude of new voices may even have made it harder to get attention. The Ministério Público has shown surprising ability to cut through the cacophony, and it will be interesting to see whether its success in taking on the powerful eventually generates some kind of backlash.

This dual reality of clear gains with ongoing limitations made the 1970s and 1980s challenging years for the environmentalists who have consistently promoted environmental protection in Brazil. The processes and outcomes that we have described were only some of the strategies they debated and tried. Through their successes and failures, environmentalists found themselves as thoroughly transformed as the politics around them. In the next two chapters we reverse the focus and look at environmentalists as central figures in Brazilian environmental politics during these decades.

2

National Environmental Activism: The Changing Terms of Engagement

In Brazil as elsewhere, what it means to be an environmentalist is forged in conflicts that occur when ideas and beliefs are played out in practices. In the process ideas change, as do practices and those practicing them. In the case of environmentalism, an especially fruitful set of conflicts arises when beliefs about the rights and responsibilities of humankind and nature become part of the same conversation as other beliefs about core identities and practices.

The formative period of groups and institutions is the crucible within which the initial responses to these challenging juxtapositions are developed. For Brazilian environmental organizations, there were three such periods in the second half of the twentieth century. The first was a period of developmentalist nationalism from the early 1950s into the early 1970s that gave birth to Brazil's oldest conservation organizations, its scientific research institutions, and the generation that created Brazil's first state environmental institutions, including people like Paulo Nogueira Neto and Maria Tereza Jorge Padua. The second was the period of political liberalization that began in 1974 and continued into the mid- to late 1980s. This period gave rise to activist organizations, whose practices developed alongside, but not always together with, a diverse set of movements that pressed for improved social conditions and a democratic political process. The third period is marked by the triple challenge of democratic restoration, economic crisis, and an exponential increase in foreign contact, the working out of which continues today (see chapter 3). We refer to these three periods as "waves" of Brazilian environmentalism.

Although the third wave saw a great increase in the amount and breadth of interaction between Brazilian and foreign environmentalists, and an increase in international experience on the part of Brazilian activists more generally, contact had been there from the beginning. Conservationists have long participated in international associations and conferences, studied abroad, and associated with expatriate scientists in Brazil. In southern Brazil, whose German heritage remains strong, the influx of environmentalist ideas was reinforced by scholarship programs and by cycles of debates and conferences on environmental topics sponsored by the German consulate in Porto Alegre and by the Association of Ex-scholarship Holders in Germany (AEBA, an association with chapters in fourteen states in Brazil). The International Rotary Clubs also took up environmental protection in the 1970s. The return of Brazilian exiles at the end of the 1970s, mainly from Europe, reinforced the circulation of ideas. Throughout this entire period there was always contact regarding the Amazon between Brazilian and foreign scientists (both physical scientists and anthropologists). Although these were not always friendly contacts, they were much more collaborative than one might imagine in light of the suspicion with which Brazilian politicians have tended to view foreign interest in the region (a subject to which we return in chapter 4). Officials from state agencies participated in international conferences and working groups on water and air pollution and other topics; international development bank missions also brought about exchanges of specialists. Thus international contact has always been present. At the same time, the focus of Brazilian environmentalists has remained fairly circumscribed by national problems, and most remain committed to finding national solutions for them.

Our discussion builds on several earlier influential classifications of Brazilian environmental movements. Maria Helena Antuniassi and her colleagues at the University of São Paulo studied members of self-described environmental organizations and their sympathizers in São Paulo in the 1980s. Antuniassi and her collaborators distinguished between conservationists and environmentalists (*ambientalistas*), to whom they attributed an organicist cosmology and a desire to spread a consciousness of participating in a web of living relations, while also taking action to remove present harm. Conservationists, by contrast, were said to be acting on the belief that human survival was threatened by the values of the predominant socioeconomic system; the challenge was thus to harness technical and scientific knowledge to generate new strategies for resource use (Antuniassi, Magdalena, and Giansanti 1989).

Eduardo Viola (1987; 1992; 1995), probably Brazil's best-known environmentalist scholar, has produced an evolving series of classifications whose categories correspond well to the times that they reflect. He first divided "new environmentalist" organizations post-1971 into two types: urban movements protesting environmental degradation and alternative rural communities. Both considered themselves apolitical. After the direct gubernatorial elections of 1982 Viola noted the spread of ecological ideas to other social movements, and in the middle class more generally. With redemocratization, activists tried to influence both public policy and the electoral process; Viola categorized them as fundamentalists, realists, ecocapitalists, and ecosocialists, on the basis of their political allegiances and alliances. Realists (or pragmatists) predominated, but there was a growing influence of ecosocialists, as environmentalist ideas penetrated sectors of the new left within the Workers' Party (PT). By 1992 Viola had moved away from looking at the environmental movement by itself and toward a multi-sectoral view of Brazilian environmentalism that recognized a variety of private and public actors (Viola and Leis 1995).

These classifications resembled those used by scholars elsewhere to categorize environmentalism and other "new social movements" that appeared to reflect "post-industrial" values more than material needs (e.g. Cohen 1985; Offe 1985). Dryzek and Lester (1989) described a spectrum that ranged from conservative environmentalists (including business groups similar to Viola's ecocapitalists), to large centrist lobbying groups, to groups seeking a new politics, "rejecting both the economic and political institutions of the status quo." Some radical environmentalists, they claimed, reject politics altogether in favor of direct action (314–15). Dobson (1990) distinguished between conservationists, reform environmentalists, and radical ecologists. What most differentiates Brazil in the context of democratization was the tendency, which Viola described but did not analyze, for activist environmentalists to also identify themselves with the political left. Dryzek (1997) later gave a broader theoretical framing to these classifications, calling them environmental discourses. Discourses contain elements essential for constructing stories. They recognize particular actors or entities, make assumptions about natural relationships, attribute motives to agents, and use certain metaphors or rhetorical devices (spaceships, organisms, the tragedy of the commons) to win attention (Dryzek 1997, 15–18; see also Keck 1995). All these distinctions help us to recognize the available constellation of ideas of environmentalism, but they do not yet take the next step and establish a relationship between these ideas and action. Understanding environmental politics requires looking beyond identity to an examination of how actors think about strategy, which entails appreciating how they relate to other constellations of ideas, other social and political actors, and the structure and dynamics of political opportunity that pertain at any given moment. Environmentalists develop action discourses and practices that may or may not fit perfectly with their overall cosmologies. These require special kinds of stories that indicate causes and signal remedies in particular contexts and debates, identified as *agitational niches* in which to insert the environmentalists' campaigns (Keck and Sikkink 1998b). In the process, they modify their own and others' perceptions of possibility, and the conditions of possibility themselves. If their campaigns are successful, they provide new framing stories with which others build their own action discourses; if they fail, they provide cautionary tales of practices to be avoided.

The First Wave: Conservation Movements

This story of conservation movements in Brazil begins in the 1950s, but actually draws on a longer history of conservation thinking in Brazil. The environmental historian José Augusto Pádua found environmental critiques dating from as early as 1786, showing that well-known Brazilian social thinkers such as José Bonifácio and Joaquim Nabuco were aware quite early of some of the environmental consequences of Brazil's patterns of development (Pádua 2002). Later on came civic conservation movements such as the Brazilian Excursionist Center, formed in Rio de Janeiro in 1919, and the Brazilian Federation for Feminine Progress, whose leader Berta Lutz was a biologist. In the early years of Brazil's modernizing president, Getúlio Vargas, such groups took advantage of "the reformist, innovative atmosphere of the moment" to influence a new Forest Code and the 1934 Constitution (Dean 1995, 259–61). But these early initiatives were soon overwhelmed by the push for development, and their successors only appeared at our starting point, the 1950s.

The most prominent of the new organizations was the Brazilian Foundation for the Conservation of Nature (FBCN) in Rio de Janeiro (then Brazil's capital). A small cluster of organizations formed in its ambit in the mountain towns around Rio as well as several in São Paulo. The FBCN began in 1958 when a group of twelve agronomists, biologists, and journalists got together to stop what they saw as the degradation of the country. Although they explicitly excluded politicians from the founding group and rejected partisanship, the organization subsequently counted congressmen among its members. It is formally registered as a civil organization in Brazil and joined the International Union for the Conservation of Nature (IUCN) when it began. Because the FBCN is both a prototype and the predominant Brazilian conservation group in this first phase, its personnel and projects are worth examining more closely.

The career of Alceu Magnannini, president of the FBCN in 1990, is typical of the group's members. Magnannini began as a biogeographer at the Brazilian Institute for Geography and Statistics (IBGE), then studied for a degree in agronomy. He worked for eight years at the Botanical Gardens in Rio and also for the government forestry department. Most FBCN members have been civil servants, journalists, or politicians. Besides attempting to introduce conservationist policies and values in professional settings, the group has undertaken conservation projects, relying on volunteers to carry them out. Financing has come from government agencies, generally as fees for service, and foreign foundations. Although they have sought it, the group's leaders have failed to win private sector support. When new conservation groups took shape in the late 1980s, competing for the same money, the FBCN suffered from a funding squeeze. By early 1989 it had only eight employees with advanced degrees, where just a few years before it had employed thirty scientists and technicians (interview with Magnannini 1990).

Formed shortly before the military coup in 1964, the FBCN did not undertake big public campaigns or large meetings with other environmentalists to discuss common ideas. In any case, for organizations like the FBCN *science* was the most important kind of activism. Advances in scientific research were essential to persuade policymakers to act, these conservationists believed, as was a prolonged effort to educate politicians and the public about the need to conserve natural areas before they were lost forever. Science was the basis for their collaborative work with the government and international funding agencies in planning and setting up national parks. Although not all the conservationists of this generation were hostile to the new kinds of environmental activism that arose in the late 1970s, they believed that their approach, grounded in expertise, was more serious and in the long run more valuable.

The FBCN and other conservation groups were formed to speak to the issue of rapid economic development and its impact on natural areas. The language of science that they spoke resonated with the technocratic and modernizing orientations of developmentalist governments of the time. Even when the military regime severely constricted political space, scientists and technical personnel retained much of their prestige with the military, which sought their imprimatur for its policies. Maria Tereza Jorge Padua, a conservationist with the IBDF, even found the military era an "anomalously favorable" time for conservation policies (Foresta 1991, 80). As the environmental activist and journalist Randau Marques put it, "The military respected those who knew how to make nuclear weapons" (interview with R. Marques 3 April 1991).[1] Thus scientists, often speaking through the Brazilian Society for the Advancement of Science (SBPC), could and did oppose some developmentalist policies, especially regarding the Amazon.

The organizations formed in the 1950s and 1960s to protect natural areas and wildlife were in the main preservationist, in the sense that they believed set-asides were the safest way to conserve the nation's environmental heritage. As late as 1972, date of the World Conference on National Parks, the FBCN president José Candido de Melo Carvalho opposed leaving indigenous people in protected areas, arguing that they were inevitably deculturated and thus became a threat to nature (Foresta 1991, 63). Arguably many of their "new environmentalist" successors have also fought for set-asides, albeit with different rhetoric, especially about indigenous populations. What distinguishes the earlier generation from the "movement" environmentalists who came later was its unequivocal faith in the advancement of science and scientific rationality as the appropriate remedy for environmental ills.

In this respect, Brazilian conservationists fit the model described by Samuel Hays for the United States: "Conservation, above all, was a scientific movement, and its role in history arises from the implications of science and technology in modern society . . . Its essence was rational planning to promote efficient development and use of all natural resources . . . It is from the vantage point of applied science, rather than of democratic protest, that one must understand the historic role of the conservation movement"

(Hays 1959, 2). There were strong affinities between Brazilian conservationism in these early stages and its counterparts in more developed regions. Despite sharp differences among later environmental movements in the United States, the United Kingdom, Germany, and Norway, all had a similar first generation of conservation movements (Dryzek, Downes, Hunold, and Schlosberg with Hernes 2003). International organizations like the IUCN, formed in 1948, helped to promote this first wave.

At the same time, Brazilian conservation movements added some original elements to the international model. Among the pioneer conservationists in Brazil, faith in science was coupled with a strong dose of developmentalist nationalism. Many of them came of age in the heady, optimistic period when President Juscelino Kubitschek (1955–60) promised to achieve fifty years of development in five (Sikkink 1991). Conservationists did not reject Kubitschek's development goals per se; instead they wanted to ensure that development plans were implemented rationally. This was the generation called upon to run the environmental institutions formed in the aftermath of the United Nations Conference on Human Development at Stockholm in 1971, beginning with the appointment of Paulo Nogueira Neto as the first head of the secretariat of the environment, which began to function precariously in 1973 (see chapter 1).

That nationalist and developmentalist ideas were shared across the political spectrum helps to explain some of the strange bedfellows who appeared in the early Brazilian conservation movement. A case in point is the National Campaign for the Defense and Development of the Amazon (Companha Nacional para a Defesa e o Desenvolvimento da Amazônia, or CNDDA; interview with Valverde 1990). As Arnt and Schwartzman wrote in their survey of organizations active in the Amazon, the CNDDA "was a bridge between the ecologism of the 1980s and the national-popular tradition that comes straight from Euclides da Cunha, the indianism of Marechal Rondon, the Prestes Column's voyage of discovery, and the work of Darcy Ribeiro. Besides the CNDDA and the Brazilian Forest Foundation, there are few groups within environmentalism that can make the link between the past and the future" (Arnt and Schwartzman 1992, 131). Founded in Rio de Janeiro in 1967 in reaction to a set of working papers published by the Hudson Institute, the think tank led by Herman Kahn that touted the advantages of building enormous dams in the Amazon basin (see Keck 2002, 47), the CNDDA brought together scientists and progressive ex-military officers (many members of the famous Brazilian Expeditionary Force during the Second World War). Some were members or sympathizers of the Brazilian Communist Party (PCB) and were among the few environmental activists to be jailed during the military period—not for being environmentalists but for being communists (interview with Muser 1991). The group's honorary president was the conservative historian Arthur Cesar Ferreira Reis, governor of Amazonas from 1964 to 1967. Reis took on the Hudson Institute in the media and in 1960 published a book called *The Brazilian Amazon and International Greed (A Amazônia Brasileira e a Cobiça Internacional)*, which traced foreign designs on the region from the seventeenth century on (Reis 1982). The book, still in print, has a conspiratorial tone that resonates well with Brazilian suspicions about the motivations of foreign environmentalists.

Shared nationalism was clearly the most important unifying factor during this first wave of environmentalism in Brazil, made up of comparatively isolated groups, separated from each other and from the larger public. Their scientific and rationalist discourses sought to amend rather than overturn the dominant political and economic logics of the time. They worked on individual projects such as preparing forest management plans or scientific research and counted on the power of their knowledge and expertise to sway the policy makers of the military era. Not surprisingly, many in this first wave felt marginalized by and skeptical of the more politicized and oppositional character of the second wave of environmental activism (Carvalho 1988; interview with R. Marques 3 April 1991).

The Second Wave: New Environmentalism

The second wave of "new environmentalism" introduced a much more direct critique of authoritarianism, blaming it for the desecration of nature and disrespect for human life and health. In this sense its proponents were moving toward a more political appropriation of environmentalism. How should we understand the political location of the new environmental groups and causes that arose in the 1970s and 1980s? Where did the groups and the issues they raised fit in relation to other questions on the political agenda at the time, and what simultaneous events affected their ability to carry out their campaigns?

The new environmentalism was introduced in Brazil toward the end of the most authoritarian period of Brazil's military regime, the government of General Emílio Garrastazú Médici (1969–74). Having issued a new and broadly applicable national security law along with Institutional Act no. 5, a decree outlawing most forms of dissent, Médici's government took a hard line on contestation early on. In 1968 labor and student movements were put down with enough force to make it clear that other challengers would meet the same fate. After six years of this repressive climate, Médici's successor General Ernesto Geisel (1974–79) began a period of gradual liberalization, initially called *distenção* (relaxation), later *abertura* (opening). Just as the later Russian glasnost led to perestroika, this liberalization opened the door to democratic transition. Although Geisel initially intended simply to release some of the pressure built up under his predecessor, the cumulative effect of his reforms went beyond that. Press censorship was gradually lifted. Continuing legislative elections became meaningful, as the quasi-official opposition party began to behave like a real opposition. New social movements and unions petitioned for attention to the high cost of living and relief from years of falling real wages. In addition, the military government agreed at the end of the 1970s to end the use of Institutional Act no. 5, restore habeas corpus and judicial independence, give amnesty to exiled opponents, and loosen the restrictions on forming political parties (Skidmore 1988; Stepan ed. 1989).

These gradual changes opened space for a flood of oppositional organization and activism in the next years. The late 1970s and early 1980s in Brazil fit the classic dynamics of a cycle of social movement protest (Hochstetler 2000), in which social movements arise within "a phase of heightened conflict and contention across the social system" (Tarrow 1994, 153). Although the first organization of the new environmentalism, AGAPAN in Rio Grande do Sul, was formed at the height of the Médici government, most second-wave groups began during the long transition to democracy (1974–89). This was a highly politicized moment, although one in which the channels for political influence were not clearly specified. Those who joined national environmental mobilizations in the second wave became politically active in two ways. Some began their activism as part of the broad new political opposition to the military and then turned to more specifically environmental activism. Others, especially in the South and Southeast, began as local environmental activists and gradually moved to larger networks and more directly political strategies to advance their environmental agendas.

The transitional period in Brazil was a period of political experimentation, in which very broad political alliances were possible. Elite dissidents joined with trade unions, artists, student organizations, and slum neighborhood committees in the name of forging a democratic opening (Alves 1985). That this democratic opening meant something different to each group was a problem lurking in the background, but one capable of being deferred, in the view of most participants, until a later date. At the end of the 1970s the formation of the PT broke this consensus by ensuring that different conceptions of democracy, and hopes for democracy, would be explicitly recognized (Keck 1992), contributing eventually to the socio-environmental current (Viana, Silva, and Diniz eds. 2001). Initially, however, insofar as environmentalists operated within a context where many different new groups and campaigns were framed as contributing to the emergence of a civil society committed to political democratization, they could gain allies from other groups contesting authorities within that frame (Hochstetler 2000). The campaign against the proposed site of a new international airport in Caucaia, São Paulo (see below), and the first campaign to protect the Amazon in 1978 (see chapter 4), were good examples of this.

When Geisel announced the beginning of the liberalization process in 1974 Brazil was nearing the end of a heady growth spurt that had brought annual growth rates exceeding 10 percent since 1968. The two oil shocks of the 1970s slowed the overheated economy, and the ensuing debt crisis of the 1980s brought it to a standstill. In the troubled context of famine after feast, it was generally easy for government officials and industrial polluters to characterize the tradeoff as one of jobs versus environmental protection, just as they have habitually done in the industrialized North. Nonetheless, Brazilian environmentalism developed and flourished in this context of economic stagnation, as its proponents demanded the right to participate in decision making and proclaimed the importance of environmental education. Thus both political and economic developments increased the proximity of environmental concerns to the democratization agenda in Brazil.

Although many of the environmental activists who came to prominence during the second wave still believed that science was central to solving ecological problems, "mainstream" science shared space with an often hostile "holistic" approach. Some of its adherents rejected altogether the notion of scientific rationality as the basis for an ecologically sound society. Ecological values were not supposed to produce an improved version of an old paradigm, but instead were the foundation for a new, earth-centered paradigm, in which human activities were radically rethought. Even those who continued to value "scientific rationality" no longer thought its arguments persuasive enough by themselves to thwart private activities or public policies that degraded the environment. That would require a long-term effort at massive public education about environmental matters, and in the short term, targeted campaigns to oppose particular abuses or undertakings.

During the early stages of the second wave, environmentalists were much more concerned with action campaigns than with building their own organizations. Most environmental groups remained small and local, relying on volunteers and on funds they had gathered among themselves or from selling posters and other paraphernalia. In some instances they formed regional associations like the APEDEMAS (Permanent Assembly of Environmental Defense Organizations) in São Paulo and Rio de Janeiro. A second category of organizations eventually financed themselves by contracting with municipal governments to do environmental education programs in schools or other public spaces, or as consultants on particular clean-up efforts. Although this funding often dried up when municipal administrations changed, it produced a kind of professionalization and was a very common form of collaboration between environmental organizations and the state. Other organizations gained in-kind subsidies from legislators at the state or municipal level: movement activists, hired as legislative assistants, were allowed to use the office photocopier, telephone, fax machine, and mailing privileges on their own time to further their movement's goals, no small benefit for organizations with no physical assets of their own.[2]

The first organization associated with the new environmentalism, one that exemplifies many of these patterns, was the Gaucho (of Rio Grande do Sul) Association for the Protection of the Natural Environment, or AGAPAN, formed in 1971 in Porto Alegre, in southern Rio Grande do Sul state. AGAPAN was conceived by a small group of men who met privately in late 1970, and then in April 1971 asked the political police to let them hold a public meeting to launch the organization (*Sobrevivência* 1986). Its first president was the agronomist José Lutzenberger, a former technical advisor of several multinational agricultural chemical firms and eventual secretary of the environment (see chapter 1). The creation of AGAPAN was suggested to Lutzenberger by American organizations like the Sierra Club and the Audubon Society and by the German organization Naturschutzring. AGAPAN itself then became a model for subsequent Brazilian organizations.

From an initial campaign against tree pruning in the city of Porto Alegre, AGAPAN moved quickly to larger projects. In 1972 it mounted a successful campaign to halt noxious emissions from the Borregaard cellulose plant, across the Guaíba River from Porto Alegre. In the following year AGAPAN was sending letters to France to protest nuclear testing, and beginning to raise the question of deforestation in the Brazilian Amazon, linking deforestation by small farmers with the lack of agrarian reform in the south of Brazil. Numerous campaigns followed, both quixotic and practical. In 1976 Lutzenberger published his Brazilian Ecological Manifesto, *Fim do Futuro?* (End of the Future?), an impassioned call to action against the degradation of nature. A later president of AGAPAN, the philosophy professor Celso Marques, called the group's role in drafting landmark legislation on agricultural chemicals for Rio Grande do Sul its most important achievement in the 1970s (interview with C. Marques 1990). All these campaigns shared a more politicized and technologically skeptical attitude toward environmental activism, providing an outlet for sectors which were uncomfortable with the status quo at the height of the military dictatorship (Marques 1986, 7).

AGAPAN waged these campaigns with minimal organizational infrastructure, as was typical of the 1970s. Its own participants donated most of the funds that came into the organization, along with some small project grants from municipalities. Members of AGAPAN contributed their own considerable professional expertise, and the group positioned its campaigns to take advantage of these resources, emphasizing environmental education and publicity. Members also gave courses on environmental topics, wrote letters to newspapers and governments, made themselves available for interviews and consultation, and occasionally organized groups to lobby decision makers. None of these activities required significant funding.

Other organizations formed around that time also concentrated on particular environmental campaigns, typically aimed at local environmental problems. But by the late 1970s campaigns began to be more closely linked and ambitious. Environmentalists developed informal networks among themselves for particular campaigns and gradually built more formal ones, some including activists from all regions of Brazil. They won the support of other actors in civil society, such as the SBPC, and sometimes of state officials. We show these changes by looking at a pair of early protest campaigns. In the first, activists at the state level in São Paulo took on the military government in a campaign against a planned new airport site. Many of the same activists then built upon the resulting network to take on the even more sensitive nuclear issue in São Paulo and Rio de Janeiro. Both campaigns consisted primarily of *blocking* activities, with environmental activists creating networks capable of mobilizing sufficient resources and influence to stop government action.

Formative Campaigns: The Anti-airport Struggle

When they recount the history of the environmental movement in São Paulo, activists most often refer to a struggle in the late 1970s over the siting of a new international airport for the metropolitan area as the moment when the movement took off (Antuniassi, Magdalena, and Giansanti 1989, 25–27; interview with Born and Oliveira 1990; interview with Paiolli 1990).[3] This conflict occurred at a moment when São Paulo's physical infrastructure was expanding and many groups in Brazilian civil society were pushing against the limits of the military's constrained liberalization. Student activism was reviving, demands for amnesty were at their height, and neighborhood movements were mobilizing against the high cost of living. All the groups were seeking greater freedoms of expression, press, and assembly, and participation in decision making. Political activists discussed the possibility of creating new political parties (Keck 1992, chapter 1). The military's gradual liberalization had allowed enough space for new organizing, but the risks associated with activism remained unpredictable.

The airport campaign pulled together politically diverse individuals and organizations in a critique of an important economic infrastructure project. It engaged bureaucratic actors and mobilized the local population. Although ordinary people still feared the consequences of speaking out openly, the support of elite scientists, professional associations, service organizations like the Rotary Club,[4] and even conservative women's organizations helped to reduce the sense of risk. The campaign experienced advances and reverses as both the military government and the emerging environmental networks struggled over the limits of the political opening.

The São Paulo Association for the Protection of Nature (APPN), which led the campaign, was formed in August 1976. The founders built a network with about a hundred members, centered on the Rotary Club of Embu, just outside the city of São Paulo. Embu was known as a cultural center, home to artists, craftspeople, and a certain countercultural spirit. The group's prime mover, the journalist Valdemar Paiolli, was a former president of the Rotary Club of Embu. Other founders included local businesspeople, environmentalists, members of a philosophical club called the Order of the Grail on Earth (Ordem do Graal na Terra), journalists, poets, writers, and painters from São Paulo city itself who came often to Embu. Subsequently the headquarters moved to Cotia, where Paolli had founded a newspaper called the *Gazeta de Cotia* (interview with Paiolli 1990). APPN's creation was inspired by an occasion in 1975 when Lutzenberger spoke at a symposium on ecology cosponsored in 1975 by the Order of the Grail on Earth and the Embu city government. He returned in the following year, and the groups maintained regular contact. The original idea had been to create a subsidiary of AGAPAN in São Paulo, but Lutzenberger discouraged the idea, considering it too centralized. The group thus founded APPN. In late 1976 APPN held a book launch for Lutzenberger's ecological manifesto *O Fim do Futuro?* in Cotia.

Rumors had been circulating for some time that the government was considering the area around Cotia for a new airport. In April 1975 the majority of Cotia's population apparently opposed the airport, fearing noise and a loss of green space (*Jornal da Tarde*, 7 April 1975). Paulo Nogueira Neto, national environmental secretary, protested that Cotia was in a protected water catchment area, and in July 1975 officials denied that a decision had been made. Then in December 1976, the government announced its intent to build the airport in an area that contained a forest reserve. This admission came after its vigorous denials were rendered moot by the journalist Randau Marques, who entered the area clandestinely with a photographer who photographed a newly constructed access road in the reserve. The photo and its accompanying article headlined the back page of the *Jornal da Tarde*. According to Marques there were many people in the state government of Paulo Egydio Martins whose civic consciousness led them to leak material. Marques used to tell his friends in government always to make copies of internal reports and distribute them to other departments, thereby increasing the likelihood of leaks (interview with R. Marques 29 September 1991).

APPN began to mobilize locally (in Cotia) and ask people from other organizations to participate in the struggle. Most were cautious, as the repressive Institutional Act no. 5, still in force, made people afraid of going to meetings and giving press interviews (interview with Paiolli 1990). So environmentalists began with a legal strategy. A month after the Ministry of Aeronautics announced the choice of sites, they filed suit (an *açao popular*, or popular action) in state court to prevent construction of any public works in the Cotia State Forest except those intended for preservation purposes (*Estado de São Paulo*, 2 December 1976; *Estado de São Paulo*, 28 January 1977). They filed the suit only after a sympathetic lawyer with family ties to the Ministry of Aeronautics assured them that the military would not respond with repression (interview with Paiolli n.d.).[5]

The process moved slowly. The secretary of the environment was asked to study how to minimize the ecological impact on the area, for which purpose it brought in the Brazilian Association of Sanitary Engineers (ABES), and in March 1977 the government expropriated the area. Caucaia was a state forest reserve, one of the few remaining forested areas on the Atlantic Plateau (Planalto Atlântico). Mobilizing people near the reserve whose land would be expropriated by the airport project was fairly rapid. APPN, the Embu Rotary Club, and intellectuals held regular meetings every other Sunday. The Catholic Church also became involved, with strong support from Bishop Mauro Morelli. The Cotia group invited Professor Nanuza Menezes from the University of São Paulo (USP) Botany Department to visit the area. Menezes, president of the São Paulo section of the Brazilian Botanical Society, brought in the geographer Aziz Ab'Saber from USP, and some municipal councilors and other politicians from São Paulo. After they began to meet regularly, Ab'Saber proposed an umbrella group called the Commission to defend the Community Heritage (CDPC) to oppose the airport plans. This was formed on 11 January 1978, at a meeting held at the Biology Department of the University of São Paulo with around three hundred conservationists present, and Paolli was elected secretary general. His alternate, Walter Lazzarini, was president of the Association of Agronomists (AEA) at the time (interview with Paiolli 1990). Other CDPC leaders were Ab' Saber, the state deputy Flávio Bierrenbach, and the botanist Nanusa

Menezes. Menezes and Ab'Saber led several trips to the reserve, she talking about botany and he about the need to establish a group of reserves (interview with Ab'Saber 29 April 1991).

By January 1978 the airport controversy was generating almost daily articles in the press. Paiolli sent a telegram to the IBDF (the federal forestry institute), asking it not to act on the government's request to deforest the area without considering the report of the São Paulo Engineering Institute that APPN was distributing, and warned that there would be widespread popular mobilization against deforestation should it occur (*Estado de São Paulo*, 3 January 1998). Menezes argued that deforestation of forty million square meters of land would be an ecological tragedy, and threw herself into the mobilization. The national secretary of the environment, Nogueira Neto, agreed with her assessment. Seeking another legal mechanism to stop the process, APPN leaders considered asking for a constitutional injunction (*mandado de segurança*) on the grounds that the project violated both the Forest Code (Código Florestal) and the Law to Protect Water Source Areas (Lei de Proteções aos Mananciais). CONDEPHAAT, the state landmarks commission, received requests from representatives of the Botanical Society, APPN, a traditionalist women's organization (the Movimento de Arrecadação Feminina),[6] and the Brazilian Society of Landscape Architects that the forest reserve should be designated a landmark. The famed landscape architect Burle Marx declared the project a crime, and telegrams of protest mounted. Local farmers began to mobilize. In the state legislature Assemblyman Antonio Carlos Mesquita introduced a bill to create the Morro Grande Forest Reserve. Although the bill was passed, the governor vetoed it, claiming that only the federal government could legislate on forests (*Estado de São Paulo*, 4 October 1978).

In response to a government statement that if the area was worth preserving it would be preserved, on 16 January the CDPC decided to invite Governor Paulo Egydio Martins to visit the area. On 18 January the court refused to hear the environmentalists' lawsuit, on the grounds that the government had not yet formally decreed the construction of the airport. The CDPC established a headquarters at the Brazilian Architects' Institute, and more than a hundred people signed onto the lawsuit. Aziz Ab'Saber protested "the politics of *faits accomplis*," saying that the state environmental secretariat ought to analyze the project. Ecological caravans began to tour the area.

On 30 January the Brazilian Institute for Forestry Development (IBDF) authorized forest clearing to begin, prompting Federal Deputy Alberto Goldman and Senator Franco Montoro, both MDB politicians from São Paulo, to come out against the airport. A Gallup poll showed that 63 percent of respondents thought the solution was to modernize Viracopos Airport (near Campinas), 17 percent wanted the Caucaia airport, and the rest chose other options (*Jornal da Tarde*, 20 February 1978). Finally, the state government could not get the support of local mayors or its own technical employees. The Consultative Council of Greater São Paulo (comprising thirty-seven mayors from the region) had been sidelined, as had the São Paulo office of the federal land institute. Most government technical staff apparently opposed the project (*Estado de São Paulo*, 22 February 1978). At last, on 24 April 1978 CONDEPHAAT (of which Ab'Saber was a member) voted unanimously to begin the process of investigation pursuant to declaring the area a common heritage site, during which time no work could be done. The project ground to a halt. In contrast to the state government, environmentalists had mobilized a surprisingly broad coalition to support their side. It clearly helped that Ab'Saber was then a member of the landmarks commission. He was one of a group that had advised Governor Paulo Egydio on an environmental program at the beginning of his administration. Oddly enough, the environmentalists in the landmarks commission won support from President Geisel, who during a trip to São Paulo told the newspapers that it was important to work with environmentalists and that the idea of a green ring around metropolitan areas was a good one (interview with Ab' Saber 29 April 1991). In the end environmentalists had support from the local population, academics, many politicians in the opposition MDB party (including the future president Fernando Henrique Cardoso), journalists, and a host of others (interview with Paiolli n.d.; Antuniassi, Magdalena, and Giansanti 1989, 25).

Antinuclear Protest

While the anti-airport campaign took shape, environmental groups continued to proliferate, became more politically and organizationally sophisticated, and developed better coordination with each other. They joined another campaign in the late 1970s that struck even closer to the heart of the military government and its aims, the Brazilian nuclear program. This campaign also showed the growing diversity of organizations concerned about the environment. Loosely coordinated student, conservation, and specifically antinuclear groups attacked the nuclear program from a variety of perspectives and with multiple strategies. They had the support of Cardinal Paulo Evaristo Arns, and of such environmental leaders as Lutzenberger (Antuniassi, Magdalena, and Giansanti 1989, 27). Like the government's own nuclear plans, the environmentalist opposition was located in both São Paulo and Rio de Janeiro states, with support from elsewhere in Brazil and from antinuclear activists abroad.

The Brazilian antinuclear movement originated among physics students in response to the announcement of a nuclear agreement between Brazil and Germany in 1975. Although drawing upon scientific expertise as a resource to challenge the military government in ways reminiscent of first-wave organizations, the scope here was much wider. Both the antinuclear campaign and the airport campaign aimed to achieve transparency and democratic decision making. In other words, the source of contention was not only the content of the decisions but also the process by which they were reached.

In 1975–76 Antônio Carlos de Oliveira (Tonhão) was president of the National Conference of Physics Students (Encontro Nacional de Estudantes de Física). He was part of a commission from the Brazilian Physics Society (SBF) that drafted a "Letter from Belo Horizonte," the first critical manifesto from Brazilian civil society in relation to the nuclear accord. The physicists then formed study groups on the viability and cost of nuclear plants in Brazil. Many physics students were already active in student politics, and a few developed a slide show and gave speeches on how nuclear energy worked, its costs, and its risks. They spoke to unions, neighborhood groups, and professional associations, and at events organized by groups of ecologists, keeping up this activity through the end of 1978 (interview with Oliveira 1992).

The antinuclear movement grew after 5 June 1980, when the federal and state governments announced plans to build two atomic plants in Peruíbe, São Paulo. The site was on the coast, in a wooded area that today is the Juréia Ecological Station. The federal government's decision to locate nuclear reactors in São Paulo was spurred by Governor Paulo Maluf's enthusiasm for high-visibility construction projects. After his inauguration in March 1979, Maluf's first major proposal was a constitutional amendment to move the state capital from São Paulo to the interior. The proposal backfired, producing a counter-movement that brought together an unusual mix of upper-class women and members of professional associations (geographers, historians, biologists, and others). Tonhão, at the time technical assistant to the PT leadership in the São Paulo state legislative assembly, coordinated this movement to oppose moving the capital, and after the amendment was defeated in early 1980, he enlisted participants in the movement to oppose the reactors. Thus the Movement against Nuclear Reactors included in its leadership the Brazilian Society of Physicists, the SBPC, professional associations of journalists, architects, and engineers, and the Movimento de Regimentação Feminina, and it won support from influential individuals. An advertising agency donated free publicity, and the cartoonist Henfil did a comic book.

The movement also attracted environmental organizations. The Juréia area already had committed defenders; beginning in 1973 a small, amateur group called Pro-Juréia had been mobilizing to defend that part of the Atlantic forest against real estate speculators. João Paulo Capobianco, an early leader of the Juréia movement, points to it as one of the very first examples in Brazil of social mobilization to protect distant areas for ecological reasons (interview with Capobianco 1990). The threat of nuclear installations in the area allowed them to build a broader blocking coalition with the antinuclear groups. Pro-Juréia continued to seek formal legal protection for the area even after the threat of a nuclear installation was over. They won support from an increasingly conservation-oriented public, coordinating a petition drive to preserve Juréia that collected 150,000 signatures in two months and running a publicity campaign using origami animals and the slogan "We need your hands to make these animals real."

The antinuclear demonstrations peaked in 1980 and 1981. Several organizations joined forces to plan a march against the reactors on 6 August 1980, Hiroshima Day, establishing an annual tradition in São Paulo that continued into the 1990s. For the second half of 1980 and the beginning of 1981, the Movement against Nuclear Reactors mobilized a broad constituency of

Paulistas (people from São Paulo) to oppose locating the reactors in the state, pointing both to cost and to risk.

Amid all this activity, fundamentally distinct visions of how to portray the stakes in the nuclear issue led to a serious rift in the antinuclear movement over the demonstration of 6 August. For Tonhão's group, the argument against nuclear power was primarily economic: one nuclear reactor would buy so many housing units, or clinics, or buses and blackboards. The question of risk, raised by the partial meltdown in March 1979 of one of the reactors at the Three Mile Island nuclear plant in Pennsylvania, was relatively distant; the deadly disasters at Chernobyl and (closer to home) Goiânia were still in the future. The Hiroshima Day marchers would make their point by denouncing federal and state government policies which made big projects like this one higher priorities than the population's social welfare.

On the other side was a coalition, centered on the Art and Ecological Thought Movement (MAPE), that wanted to see the demonstration represent ecological values. It proposed a silent march, with participants wearing black armbands and carrying candles, and no speeches. This appeal for the moral high ground was a common position among environmentalists on the nuclear issue. In Rio Grande do Sul, AGAPAN mounted a major campaign against nuclear power stations as a matter of principle and as a critique of a "pattern of civilization," and even managed to get the state legislature to approve a constitutional amendment prohibiting the construction of nuclear power stations in the state (Viola 1988, 218). Antinuclear activists in Rio de Janeiro also took a strongly moral oppositional stance.

Thus the Hiroshima Day demonstration as it finally took place in São Paulo was a compromise among the various groups involved. The state legislative assembly set up a commission of inquiry into the potential economic and social effects of the proposed reactors, producing a long report. By the end of 1980, however, it was beginning to appear that the reactors were not going to be built after all, though not because the movement had grown strong enough to stop them (interview with Barros 1991; interview with Oliveira 1992; CEDI, DEPSET-CUT, et al. 1988). Little by little, the debt crisis and other considerations led the federal government to abandon most of the nuclear programs it had so exuberantly announced a mere half-decade earlier. By 1981 the expected cost of the eight reactors to be built under the accord with Germany had risen from $10 billion to $25–30 billion, and the projected cost per watt of electricity produced had risen even more (Conca 1997, 195). The Brazilian government scaled back its nuclear ambitions considerably as a result.

Although nuclear energy was thus sidelined, the military had also begun a parallel program of nuclear research and development for military purposes around 1979. Parts of it continued under wraps, including the navy's secret pilot project for uranium enrichment, housed first at the Institute for Energy and Nuclear Research (IPEN) at the University of São Paulo, and later at the navy's Aramar facility near Sorocaba (Conca 1997, 195). In 1987, prodded by state legislators in São Paulo and a very broad-based local coalition in Sorocaba, the navy admitted the program's existence. The movement against Aramar mobilized a range of organizations from the Rotary Club to trade unions and student groups. In a city of 400,000, it mounted a demonstration of 30,000, and other towns in the region mobilized as well. But after the navy's admission, and its official inauguration of the facility in 1988, the movement went into decline, and the research facility continued to function. This set of events received little attention in the national press but served as an impetus for the formation of an environmental movement in Sorocaba (interview with Barros 1991). In 1991 the Brazilian military essentially traded in its desire to produce a bomb for the ability to maintain control over the nuclear program. In 1991 Brazil and Argentina, both of which had previously refused to ratify the Treaty of Tlatelolco (1967) banning nuclear weapons in Latin America, established a joint nuclear safeguard agency and concluded an agreement with the International Atomic Energy Agency (Conca 1997, 202–3).

From Protest to Engagement

The large blocking campaigns against the airport and the nuclear program brought quite different kinds of activists together—and also pulled them apart as they discovered their differences. Although there were several important organizations already in existence, the APPN and especially the CDPC became the base of the family tree out of which grew most of São Paulo's environmental movement. In the early 1980s two vectors emerged. One united many of the existing groups, including the Grupo Seiva de Ecologia (led by the actress Cacilda Lanuza), Ecological Union (União Ecológica), and Ecological Action (Ação Ecológica), into a Permanent Assembly of Environmental Defense Organizations (APEDEMA). APEDEMA's initial leaders included Walter Lazzarini, elected in 1982 to the state assembly and later a regular presence in government environmental agencies. The other vector was oriented around a newly created organization, OIKOS, which arose at the end of 1981. The divisions between the two, according to Rubens Born, were political, though not partisan. They concerned questions like the role of a nongovernmental organization under an authoritarian regime, as well as strategic questions about alliances, targets, and results (interview with Born 1990). These questions were especially pressing during the uncertain transition period, and sharply different answers led for a while to deteriorating personal relations among environmental activists. Most of the CDPC's leaders left it. Those who valued a continuing oppositional, anti-political stance and decentralized organizational forms organized the APEDEMA (interview with Lanuza 1990).

OIKOS offered a different set of answers, prefiguring in many ways the professionalization of organizational structures that took place later in the decade, during the third wave. OIKOS was made up of university graduates, liberal professionals, and businessmen. Founders included Fábio Feldmann, a lawyer who started the environmental and human rights commission of the São Paulo branch of the Brazilian bar association (Ordem dos Advogados do Brasil, or OAB), Randau Marques, the city's best-known environmental journalist, the businessman Roberto Klabín, and Rodrigo Mesquita of the Estado de São Paulo newspaper group. OIKOS was intended to put a more professional gloss on environmental protest. Its approach was to denounce government actions and the development policies that had produced the environmental problems in the first place. To do this it sponsored annual campaigns that borrowed techniques from the public relations and advertising fields: the campaign in 1982 was about the Pantanal wetland, in 1983 about pesticides sprayed aerially in cotton plantation areas in São Paulo, and subsequently about industrial pollution in Cubatão. OIKOS was especially active in the Cubatão case (discussed further in chapter 5). Feldmann was the attorney for the Cubatão Victims' Association, helping it to obtain legal status, and Randau Marques's articles on Cubatão in *Jornal da Tarde* amounted to a crusade.

In these ways OIKOS's members followed the earlier environmentalists' strategy of using their professional qualifications in the service of their activism. But in a more politicized age, the relevant skills expanded beyond scientific ones. Eventually they included Feldmann's political skills as a federal deputy and later state secretary of the environment for São Paulo. With Feldmann's election in 1986, OIKOS became something of a political lobby for the Constituent Assembly (see chapter 1). Especially after Feldmann's entry into politics, some environmentalists criticized OIKOS as too closely tied to his political career and party (interview with Lanuza 1990).

Meanwhile in Rio de Janeiro, the nuclear issue and protests also pushed local environmentalists in new directions, albeit different ones from those of São Paulo. The Angra I nuclear plant there, which the Brazilian diplomat Saraiva Guerreiro called *"essa infeliz usina"* (that unfortunate power plant) (Guerreiro 1992, 150), was not a product of the nuclear accord with Germany. In 1968 the national nuclear energy commission signed an agreement with Eletrobrás, through its subsidiary Furnas, for construction of a nuclear reactor in Angra dos Reis. The enriched uranium plant was built by Westinghouse on a turnkey agreement, with no technology transfer. Despite these limitations Angra remained a part of an overall governmental strategy, dating from the 1950s, of trying to achieve autonomy in the nuclear area. From the beginning the plant was full of defects, a fact that only highlighted some of its logistical failings (lack of an evacuation plan in case of accident for the nearby population) and its improper siting (near a geological fault, on the beach of Itaorna, which in indigenous language means "rotten stone") (CEDI, DEPSET-CUT, et al. 1988, 23–24). In 1982 a group of environmental and antinuclear activists began to organize a series of actions in Angra dos Reis under the name of "Hiroshima, never again" (Hiroshima, nunca mais). They filed an *ação popular* in federal court that resulted in the closing down of the reactor in March 1986, and brought in technical experts to assess security there (Arnt and Schwartzman 1992, 325). Antinuclear activism aimed at Angra raised the profile of activists in Rio de Janeiro who would later form the Green Party, and helped to turn the local debate over whether such a party was apposite into a national one. After protesting and sometimes blocking the military government's initiatives by means of such actions, environmental activists found themselves in quite a changed position a decade after AGAPAN's founding. Although most organizations continued to be small and voluntary, with minimal funding, they were now more closely linked to each other and had developed considerable mobilizing experience by using a variety of strategies. As Brazil continued to return slowly to civilian liberal democracy, environmentalists began to think about engagement rather than confrontation in the political realm, which they conceived of as both a set of formal institutions and a public sphere of debate. Their protest campaigns had produced some leaders who advocated participation in party politics. Others wanted to remain outside formal politics. In the mid-1980s these choices produced extensive debates in the movement about identity and strategy.

The Problem of Political Participation: Party Politics

The 1980s were years of political democratization in Brazil, beginning with the creation of new political parties in 1980 and ending with direct presidential elections in 1989. In 1982 state governors were elected directly for the first time since 1965, along with state assemblies, municipal councils, and most of the federal legislature. By mid-decade press censorship had almost disappeared. Although the millions of people who took to the streets in 1984 to demand immediate direct presidential elections failed in their campaign, indirect elections in the following year nonetheless brought a civilian to the presidency and effectively ended military rule. Legislative elections in 1986 took on higher than usual salience, as the combined Chamber of Deputies and Senate were to act as a National Constituent Assembly, charged with writing a new constitution for a democratic Brazil.

Party politics in the mid-1980s were volatile. Unlike military dictatorships elsewhere, the one in Brazil held regular if highly controlled legislative elections between a pro-government party (ARENA) and a legal opposition (MDB). When the government first allowed the opposition party to campaign on television in 1974, the MDB made huge gains, and in subsequent elections the military had to resort to gimmicks (*casuísmos*) to retain control. With a new party law in 1979, military leaders hoped to keep the pro-regime coalition intact while the opposition splintered. By 1982 Brazil had five legal parties, of which by far the largest were the successor parties to ARENA (called the PDS) and the MDB (called the PMDB) (Keck 1992; Mainwaring 1991). The novelty was the PT, product of a confluence of social movements, labor activism, grassroots Catholic organizing, and opposition intellectuals frustrated with the reproduction of the same elitist political structures that had always prevailed (Keck 1989; Keck 1992; Branford and Kucinski with Wainwright 2003; Nylen 2003). Subsequently the configuration of parties changed with each election, and popular frustration with them grew apace. Nonetheless, in the run-up to the Constituent Assembly, domestic electoral politics was the privileged sphere of democratic organizing.

Environmentalists thus had to decide on the form that their participation in the democratization process would take, as they moved from protest to active political engagement. There were difficult choices ahead about how to relate to other social movements, the state, political parties, private firms, environmentalists abroad, and domestic and international funding sources, and indeed how to carry out the process of differentiation among themselves. Most environmental organizations were local; information flows among them were still embryonic in the mid-1980s, and contact with organizations outside Brazil was rarer still. The ease of information exchange that became commonplace with the spread of electronic media over the next decade was still unimaginable, making face-to-face meetings all the more important. In a variety of encounters, movement activists engaged in heated debate over these choices, which subsequently influenced their organizations' structures and practices.

The first political openings resulted from preparations for the constitution-writing process of the mid-1980s. Environmentalists held a number of encounters to discuss what political role they could play. Rio Grande do Sul and São Paulo held state-level meetings of environmental activists in 1984, followed by Santa Catarina, Paraná, and Rio de Janeiro (all southern and southeastern states). Many participants, mistrusting professional politicians, wanted to see a Constituent Assembly separate from the National Congress, in which a portion of delegates would be chosen without party affiliation to represent particular causes. In São Paulo the sixty groups attending a state "Conference of Ecologists and Pacifists" in April 1985 espoused nonpartisan participation in the Constituent Assembly (Nogueira and Carvalho 1985, 86). Optimists believed that they could win as many as twenty-five seats in this manner, or almost 5 percent of the total (*Jornal do Brasil*, 26 January 1986). That hope was dashed in November 1985, when Congress passed Constitutional Amendment no. 26, establishing that elections to the Constituent Assembly, like other legislative contests, would require candidates to run on party slates (Martínez-Lara 1996, 57–59). Even so, the Constituent Assembly still provided an opportunity to draw up a platform of shared principles for national debate.[7]

For many environmentalists, the vehement debates about whether or how to participate in the Constituent Assembly process offered a political education regarding electoral and formal politics (Antuniassi, Magdalena, and Giansanti 1989, 52). Should environmentalists throw in their lot with the PT's challenge to the status quo, or the catchall PMDB? Form a green party in which the environment would always be top priority, as antinuclear activists in Rio de Janeiro proposed? Or remain "above parties" altogether, trying to persuade existing parties to adopt green platforms? Embedded in these choices were still others, more controversial, but pressed hard by those who saw in "ecologism" an alternative cultural vision encompassing feminism, gay rights, the legalization of marijuana, and a variety of other so-called lifestyle issues. Debates over these questions came to a head in the mid-1980s as all Brazilians were challenged to translate their principles into new political practices.

With the discussions of electoral politics, the center of gravity of Brazilian environmentalism moved definitively from the historic first wave of conservation groups to the more politicized second wave. Conservation organizations, led by the FBCN, met and espoused a constitutional right to a sound and ecologically balanced environment, and the responsibility of citizens to defend it (*IV Encontro Nacional* 1984). At their next conference two years later, conservationists approved specific language for such provisions (*V Encontro das Organizações* 1985). Second-wave environmentalists agreed with these constitutional ideas but preferred to promote them through more partisan strategies—although they became deeply divided between two partisan choices. One faction, led by returned exiles who had been inspired by the Green Party politics they had encountered during their European sojourns, wanted to establish a Green Party in Brazil; another wanted to promote environmentalist ideas in all political parties. The two groups eventually agreed to disagree, resulting in two modes of political engagement after 1985. Some formed a Green Party, and the others worked through an Interstate Ecological Coordination for the Constituent Assembly (CIEC) to create a "Green List" of candidates.[8] Both party-based strategies crowded out the conservation groups' more expertise-based strategies. Some historic conservationists resented the greater visibility of more recently established second-wave groups, and were uncomfortable with what they saw as their "leftism" and "arrogance" (Carvalho 1988).

The Green Party Option

The Green Party proposal began to take shape in June 1985, in discussions among antinuclear and environmental activists in Rio de Janeiro. In November they invited counterparts from other states to join in.[9] Although many remained skeptical, the core group started to organize in Rio at the end of 1985 and officially founded the Brazilian Green Party (PV) on 18 January 1987. Several key leaders in this group had embraced the ecological cause and countercultural repertoires of European Greens, finding in them a cultural, spiritual, and political critique both of the status quo and of the traditional left. Fernando Gabeira, Alfredo Sirkis (also spelled Syrkis), Liszt Vieira, and Carlos Minc returned from exile after the Amnesty Law of 1979 was enacted, already determined to create a Green Party in Brazil (Pádua 1992, 139). Soon after their return, both Gabeira and Sirkis published best-selling books based on their guerilla experiences (Gabeira 1979; Sirkis 1980). Vieira was elected to the Rio state assembly on the PT ticket in 1982 but soon wanted to be in a party that had a clear environmental identity (interview with Vieira 1991).

Gabeira and the others bided their time during the first rush of party formation in 1980–82. The transition's uncertain trajectory made most opposition actors hesitant to abandon the PMDB, successor to the MDB as the big tent opposition party. Apart from the Workers' Party, which went to some lengths to forge an autonomous identity, opposition parties of the period were personalistic vehicles. Some leading environmentalists like Lutzenberger of AGAPAN opposed linking their struggle to any party at all, arguing that environmentalism had an "identity much too broad for most political ideologies."[10] The future Greens originally wanted to set up green caucuses inside all the opposition parties, especially the PMDB and the PT (Sirkis 1996, 199), and created a statewide Green Collective in Rio de Janeiro. Yet five years of working within the left and center-left convinced them that these parties remained hostile to ecological ideas, to say nothing of ideas about feminism, homosexuality, and alternative lifestyles (interview with Sirkis 1992).

These "amnesty environmentalists," as Sirkis (1996, 200) called them, did not deny the PT's argument that Brazil critically needed to combat class-based oppression, but they were equally determined to fight oppression based on gender, race, sexual preferences, and lifestyle.[11] PV founders still unequivocally located themselves on the left of the political spectrum. The amnesty environmentalists had cut their political teeth on the left, in Marxist guerilla organizations and subsequently in legal leftist parties. Gabeira called Brazil's striking juxtaposition of poverty and industrial development *poliséria*, combining the Portuguese words for pollution and misery (Gabeira 1987b, 172). The party's founding manifesto takes on "oppression, inequality, hunger, misery, elitism, corruption, cultural backwardness, and the remnants of authoritarianism." Moreover, in their Action Program, the Greens called for a society in which democracy prevailed not only in representative institutions but also in daily practices—the workplace, the justice system, and energy and technology policies—and for all Brazilians, including indigenous peoples, women, minors, Afro-Brazilians, and gays and lesbians. These values resonated with the experience of Brazilian social movements opposing the military government, as well as the banners of European Green parties (Mainwaring and Viola 1984; Spretnak and Capra 1986). In their bylaws, the Greens committed the party to internally democratic practices, setting high ethical standards for members' behavior. Like Green party members elsewhere, they blended libertarian and participatory political values with conservationist aims, eventually calling these the values of "self-governing ecological socialism" (Partido Verde 1986).[12]

The prospect of the Constituent Assembly spurred the group to action. Constitutional Amendment no. 25, enacted in May 1985, had lowered some of the hurdles to party formation, authorized split tickets, and provided free radio and television time for all legal parties, regardless of whether they currently held legislative seats (Martínez-Lara 1996, 46; Mainwaring 1991). Whatever the immediate chance of election, media access gave them a platform from which to expound their ideas. Because the party reforms retained proportional representation with nominal voting, multiple parties, and no threshold, their chances were better than they might have been otherwise (Lamounier 1989b, 115). Even so, forming a party was still not easy. The bureaucratic minutiae required for legal registration made for an uncomfortable fit with the PV style, and indeed the party gained and lost its legal registry several times before achieving definitive registration in 1993. PV activists wanted to encourage local-level deliberative democracy, their goal being a party that was also a movement and that encouraged participatory democracy in a wide range of spheres (Gabeira 1987b). Rhetorically speaking, these resembled the participatory goals that the PT had also advanced four years earlier.

The Greens' rhetoric notwithstanding, many environmentalists still suspected them of seeking to monopolize the political expression of environmental causes. Moreover, the authoritarian personal styles of party organizers cast doubt on their commitment to participatory democracy. Although the press hailed Gabeira and his colleagues as the country's "best-known ecologists," in the environmental movement they were latecomers to the environmental cause, with little experience in grassroots environmental activism in Brazil. In 1985 Caio Lustoso of AGAPAN, then a city councilor, commented disdainfully that "with rare exceptions, the ecological movement in Rio is not very strong . . . and rarely undertake[s] any ecologically important struggles" (*Isto É*, 4 December 1985, 80). Some in São Paulo characterized the Greens as the "festive" wing of Brazilian environmentalism.

In summary, in 1985–86 Green Party advocates proposed a particular kind of environmental party, one that they saw as embodying two innovations: it was to be organized as both party and movement, and it was to have a particular vision of self-governing ecological socialism. Opponents argued in response that a Green party would ghettoize environmental concerns (interview with Arnt 1990) and distract from the more important task of building the ecological movement (Viola 1987). Fábio Feldmann, the only environmentalist candidate actually elected to the Constituent Assembly, feared that identifying environmental goals with a weak party would undermine the movement's influence (*Folha de São Paulo*, 4 January 1987).

There were other environmental activists, especially PT founders and candidates, who objected to the Greens' vision of "self-governing ecological socialism," and particularly its countercultural bent. While agreeing that existing socialist countries had a dismal record on the environment, they nonetheless maintained that struggle among class forces remained central to resolving environmental problems (interview with Gregol 1990). Conservationists, intent on combating the proximate causes of environmental degradation, saw no need for an ecopolitical vision at all, and accused the Rio Greens of creating confusion by bringing in issues like legalizing marijuana and other drugs, the rights of homosexuals, and so forth (interview with Guttemberg n.d.).[13] They too favored either working with other parties or lobbying from a position outside party politics altogether (Sirkis 1996, 201, 199; interview with De Barros 1990).

Finally, even some Green sympathizers objected to the timing.[14] The demobilizing effect of the military regime, Brazilians' continuing mistrust of politics, and the country's lack of experience with democracy of any kind, let alone participatory democracy, might cause most people to see the Green proposal as yet another imported idea that didn't belong—an *"ideia fora do lugar,"* as Roberto Schwartz wrote about liberalism. Founders of the European Green parties had built upon an associational base whose Brazilian counterpart was minuscule. The pacifist movement was tiny, the antinuclear movement involved many of the same people as the environmental movement, feminists were not strongly party-oriented, and Afro-Brazilians, where organized, supported either the PDT or the PT.

In the end, adherents of the different positions agreed to disagree, consistent with the environmental movement's distaste for centralized decision making. At a national meeting in November 1985, the majority agreed that Green Party supporters should carry on, provided that they did not weaken other parts of the environmental network and its projects in the process.[15] Other environmentalists would continue to seek influence in all parties. The diversity of positions on political participation were influenced not only by different ideas but also by strikingly different opportunities for political action among states and municipalities. These conditions often diverged from those found at the national level. All Brazilian political parties were at the time under ten years old and unevenly dispersed geographically. To this day most are not programmatic parties, and party elites in different states often put quite different stamps on the same party. As state governors have enormous influence over the state's congressional delegation (of whatever party), parties renegotiate their identities continuously, and party discipline rarely obtains (Mainwaring 1991). As a result, activists from different states in the 1980s drew different conclusions about how open existing parties were to the environmental agenda. In Rio Grande do Sul, for example, environmentalists found ample space for their ideas within the PT, and some long-time environmentalists were prominent in the state-level party organization (interviews with Gregol 1990; C. Marques 1990). By contrast, activists in Rio de Janeiro met resistance to their agenda in the PT and other parties there (Sirkis 1996). Keck (1992, 100ff) notes that the Rio PT was unusually divided over the coexistence of "quantitative" concerns (traditional working-class politics) and "qualitative" concerns (libertarian, feminist, and ecological). São Paulo staked out a middle ground, with some openness in the PT, the PMDB, and the PSB to environmental candidacies. Thus it should not surprise us that the strongest push for a Green Party came from Rio de Janeiro, while Rio Grande do Sul led the opposition.

The PV was officially founded on 18 January 1987, led by the amnesty environmentalists from Rio de Janeiro. It remains to this day the only Brazilian political party seriously committed to an environmental platform, despite efforts by environmentalists in other parties. Rio de Janeiro remains its strongest section. Although members of the Rio group considered themselves on the left, the profile of the Green Party outside Rio was much more diverse. The Greens (and some of its individual leaders) have had a love-hate relationship with the Workers' Party. In 1986 Green candidates ran on the PT slate in several states, and the PT considered Gabeira as a vice-presidential running mate for its candidate Lula in 1989. Nonetheless, the post-materialist values which party founders espoused clashed with the decidedly materialist orientation of the Brazilian left, even for the democratic socialists in the PT who supported environmentalism and took stands alongside feminists and other minorities. Eventually some of the founders in Rio, such as Carlos Minc, returned to the PT, on the grounds that the PV was too closely identified with a small group of people. Even Fernando Gabeira, clearly a member of that small group, ran as a PT candidate several times, beginning with his candidacy for state governor of Rio in 1986. The PV was part of the governing coalition of the first national PT administration which took office in 2003, but withdrew from the government in May 2005, after a series of anti-environmental initiatives by the PT government.

The "Green List" option

Environmentalists who did not support creation of a Green party created the cross-partisan CIEC and put together a multi-party green list (*lista verde*) of candidates for the constituent assembly. The lists were compiled in 1986 either by state-level environmental coalitions such as the APEDEMAS, or by broadly representative state meetings; they could include candidates from any party, provided that the candidate met the agreed-upon criteria. In six states green lists included candidates for federal and state legislatures, while in four others they endorsed only state assembly candidates (CIEC 1986c).[16]

Selection criteria, established by the sixty participants at CIEC's third meeting in May 1986, were based on personal histories and positions on various issues (CIEC 1986a). First, candidates were required to be environmental movement activists, or if none were running, dependable allies. Second, they had to endorse publicly a rather catholic "Ecologist Platform," product of the CIEC meeting in September 1986. The platform had eighteen points, briefly stated, beginning with "No nukes!" and ending with "For the construction of an egalitarian, just, pacific and ecologically viable society, which respects future generations" (CIEC 1986b). In between were planks that reflected the environmental movement's ties to pacifism, such as a call for converting the arms industry to peaceful uses and ending obligatory military service and nuclear production. Others called for decentralization, greater access to information, and plebiscites on high-impact development projects. There were also more directly environmental measures related to cities, workplaces, ecosystems, air, water, soil, food, and indigenous nations. The platform endorsed some of the key demands of popular and labor movements: agrarian reform, a shorter workweek, and full employment guarantees, and finally called for an end to racial, sexual, religious, and ideological discrimination. This eclectic document epitomized environmentalists' commitment to paint their movement in all its diversity.

Not surprisingly, the final green list drew heavily from the parties on the left and center-left: of twenty candidates on the lists for national office, eleven were from the PT (CIEC 1986b).[17] These included three eventual members of the PV—Gabeira, Minc, and Vieira—since the PV could not yet run its own candidates. The catch-all opposition party, the PMDB, was a distant second to the PT, with three candidates on the list. No candidates from the successor to the pro-military party were on the list. The PT and PMDB also dominated the state lists. Environmentalists hoped that having green list candidates refer to the program and the list in public debates and media appearances would by itself contribute enough to public environmental education to justify the effort. But they also expected many candidates to win. Environmentalists scheduled a series of demonstrations to build up grassroots mobilization before the November elections: on 6 August, anniversary of the bombing of Hiroshima; on 21 September, Arbor Day, to call for protection of the Amazon and other regional ecosystems; on 16 October, World Food Day, because of the environmentalists' opposition to pesticides and support for ecological agrarian reform; and on 25–26 October, when a Latin American International Eco-pacifist Demonstration was staged on the Paraguayan border. In Rio de Janeiro the amnesty environmentalists brought crowds of fifty thousand to create a dramatic human chain embracing the polluted Rodrigo de Freitas Lake, and to take part in another demonstration for women's rights (Gabeira 1987a).

Despite its initial success, the national CIEC organization crumbled in the months preceding the election, victim of intergroup rivalries and inexperience in large-scale organizational efforts (interviews with De Barros 1990; Petrillo 1990). Logistical problems abounded, with little time and less money available for the frequent meetings, travel, photocopies, and phone calls. The CIEC gathering in Rio de Janeiro in September was its last. The election results were disappointing. Running on the PT ticket, Gabeira won only 7.8 percent of the vote for governor of Rio de Janeiro, still a respectable showing given the PT's weakness in the state. Yet only one candidate from the green list, Feldmann of São Paulo, won election to the National Constituent Assembly. While Feldmann blamed the green list's poor showing on the lack of opportunity to explain to the electorate what it meant to be an ecological deputy (interview with Feldmann 1991), other problems had much more electoral salience in 1986. As the Brazilian economy moved further into hyperinflation, the temporary success of President Sarney's Cruzado Plan, meant to stabilize inflation and promote economic growth, became the most important issue in the campaign, prompting many voters to support candidates of the governing coalition and crowding out even the "constituent" nature of the Congress being elected (Martínez-Lara 1996); the environment was still further behind. These factors helped to account for the disappointing turnout. As we saw in chapter 1, environmentalists eventually regrouped behind their one electoral success and gained significant achievements in the constitutional text anyway.

Brazilian environmental activism underwent a remarkable transformation from the 1950s to the mid-1980s. A wave of scientific and expertise-grounded organizations was gradually joined by a second wave of much more explicitly political organizations. From isolated and sporadic action, environmentalists began to build both temporary and permanent networks that linked them to each other and to the larger political debates of the time. The largest context for these changes was the political transition from a developmentalist military government to the early stages of a new civilian government. In Brazil this transition was unusually protracted and gradual, allowing environmentalists to experiment with new forms of political participation and to take graduated steps toward political engagement as the political system itself opened.

In broad terms, the changes of Brazilian environmentalism fit global patterns among northern countries. All of them share this transition from an early generation of conservation groups to a more politicized generation of "new environmentalism." The Brazilian variant has its own characteristics, however. The conservation stage was joined with unusual levels of nationalism, especially around Brazil's largest ecosystem, the Amazon. Brazilian new environmentalists were much more clearly on the political left than in many other countries. That the military government's economic policies had excluded many Brazilians from the nation's economic "miracle" contributed to this result. In addition, the context of democratization gave many Brazilian environmentalists an unusual focus on questions of process and political participation, valued in themselves regardless of substantive outcomes. In the next wave, it also brought them into unusually close relations with the state.

From Protest to Project:
The Third Wave of Environmental Activism

Developments in the Brazilian environmental movement in the late 1980s and early 1990s had effects that continued to be felt into the twenty-first century. Unlike the second wave of environmental activism, this third wave did not stem from a generational change in visions and life experiences. Instead, many of the second-wave actors made new political and organizational choices about how to be environmental activists, as individuals and as organizations. Brazil entered the 1980s under military rule and exited with a democratic constitution and the first directly elected president in more than twenty-five years. By contrast, there was more continuity and considerably less to celebrate in economic conditions, including inflation, debt crises, and economic stagnation. Environmentalists believed that the new context required them to develop more sustained and positive projects for environmental protection, rather than only block the projects of others. Liberal democracy allowed these kinds of projects; the economic problems required them.

From the mid-1980s to the present, relationships between state and movements, between the private sector and public interest organizations, and among differing conceptions of environmental activity have all been negotiated and renegotiated. As new participatory opportunities appeared, some organizations wanted to aim higher than could be done through purely voluntary efforts, and chose professionalization as the best course. It was easier said than done, and stimulated new debates: Where would the money come from to scale up and form professional organizations, with offices and staff and projects to pursue? What were the advantages and disadvantages of seeking funding from members' dues, from private firms, from government sources, or from foreign governments and foundations? Was there room in the environmental movement for professional and grassroots organizations, and how would each find its niche? These questions dominated many environmental encounters during the transition to the third wave (roughly 1986–92), sometimes at the expense of substantive discussions.

The Brazilian debates and organizational transition followed on the heels of similar developments in the United States, where external funding for social movement organizations became increasingly important in the 1960s and 1970s. Scholars of these movements have concluded that the external funding created a "dynamic of financial steering" (Brulle 2000, 84) that pushed environmental organizations to less radical environmental agendas than citizen values would support (Brulle 2000, 256–58; see also Tokar 1997). Using more partial data—Brazilian organizations have no government-mandated public accounting—we examine the extent to which similar results have also appeared in Brazil.

Socio-environmental discourse developed out of efforts to work with other kinds of social movements, efforts which had rarely been made before the mid-1980s. Although at the end of the dictatorship most environmentalists considered themselves part of a broad movement for democracy and social justice, many participants in the myriad social movements demanding material improvements as well as political democratization considered the environmental message extraneous to their concerns (Stepan 1989). Bridges between environmental and other social movements were personal, not organizational. Nonetheless, this confluence in timing helps to explain why many Brazilian environmentalists were more likely to identify with the political left than were their counterparts in the North; espousal of nonmaterial values did not obviate attention to poverty, inequality, and the structural conditions of dependent capitalist development. Socio-environmentalism, which linked environmental degradation and social justice, was the frame that environmentalists chose to reach out to those who continued to be economically and politically excluded. To influence democratic decision making in a time of economic crisis, environmentalists needed to broaden the social bases of their appeal.

Almost as soon as environmentalists made the leap from local to national politics, they were pulled into engagement with the international community as well. Some conservation organizations and individual environmentalists had long-standing international ties, but most environmental groups did not. Thus in 1988–89, when environmental organizations in the United

States and Europe loudly condemned Brazil's apparent complacency as deforestation in the Amazon forest accelerated, Brazilian environmentalists knew very little about the foreign organizations, and many doubted their motives. Over the next few years, engagement with foreign environmentalists, environmental ideas, and international institutions increased rapidly. International organizations like Greenpeace and the World Wildlife Fund (WWF) organized branch offices in Brazil, attracting Brazilians already active in the environmental movement. Electronic mail became available through the Alternex system, run out of IBASE (Brazilian Institute of Social and Economic Analyses), an NGO in Rio de Janeiro; the Internet was not far behind. Most influential of all, in 1992 Rio de Janeiro hosted the United Nations Conference on Environment and Development (also known as the Earth Summit), preparation for which engaged the national environmental community for close to two years and entailed a far-reaching negotiation of ideas and identities among national environmental organizations and other nongovernmental actors in Brazil. By the end of this conference, the choices of professionalization and socio-environmentalism were well established, if never universal.

Professionalization

Other nongovernmental sectors developed professional organizations earlier than environmentalists did, especially those for social assistance to Brazil's poor (Landim 1998; Fernandes 1994). Many NGO cadres now in their fifties came of age in the popular education projects that spread throughout Brazil in the early 1960s, projects based on Paulo Freire's pedagogy of the oppressed that inspired a lasting commitment to grassroots political action. Several of the older NGOs were created with the support of the Catholic church, which also played a crucial role supporting grassroots organizations of various kinds during the military dictatorship (Della Cava 1989).[1] Some activists joined armed leftist organizations at the end of the 1960s and went into exile, mostly in Europe, with a few in Canada and fewer still in the United States. Upon their return to Brazil, some former exiles decided to create professional organizations and think tanks (Landim 1998). Academic research institutes set up by opposition intellectuals who stayed in Brazil during the dictatorship also prepared the way for social action NGOs during the transition. A number of the older organizations have transformed themselves several times. For example, FASE (Federation of Organs for Social and Educational Assistance) began in 1961 as a Catholic relief agency, then turned to community and worker development, followed by popular education and a variety of activities in assistance (*assessoria*) to unions and social movements, including environmental movements (Landim and Cotrim 1996, 126–29).

The activists at the forefront of professionalization drew on international contacts and financing to found organizations that did "action research," providing assistance to grassroots initiatives and undertaking research geared toward their needs. This included people like Herbert "Betinho" de Souza, Carlos Afonso, and Marcos Arruda in Rio, who formed IBASE, and Jean Marc Von Der Weid, who set up FASE's appropriate technology project. Afonso, through IBASE, introduced computer networking and e-mail to Brazilian activist circles in the 1980s by creating Alternex, linked to the Institute for Progressive Communication (whose affiliates in the United States included Peacenet and Econet). Afonso helped to facilitate such connections in other Latin American countries as well (Fico 1999), well before the Internet made computer connections generally available (interview with Afonso and Sousa 1992; interview with Von Der Weid 1992).[2]

This kind of public interest organization was a new development in Brazil, as it was in other Latin American countries over approximately the same period. The new NGOs produced and distributed information, and engaged in lobbying, advocacy, popular education, and sometimes the direct provision of services. In 1991 the Brazilian Association of NGOs (ABONG) was formed, engaging over two hundred NGOs in a variety of efforts from lobbying for legislation favorable to the NGO sector to joining in street mobilizations against violence (Friedman and Hochstetler 2002). Resources for professionalization were a constant concern as the NGO sector developed, and recipients were often hesitant to share information about sources they had found. In 1986 an early survey of 1,041 NGOs made questions about budgets only optional, and the survey researchers observed that asking detailed questions about funding sources was "a delicate subject, provoking somewhat vague responses" (Landim 1988, 47). ABONG's members were more forthcoming a few years later, in 1993; a plurality (44.5 percent) reported an annual budget under $100,000, with another 43.7 percent reporting budgets between $100,001 and 500,000. Of 124 respondents, only one had a budget above $3 million (Landim and Cotrim 1996, xvii). The respondents still did not provide detailed information about specific sources and quantities.

Funding for professional NGOs in Brazil is a continuing problem. The absence of a tradition of philanthropy, whether from wealthy individuals or businesses, makes it hard to sustain public interest organizations (Roberts 1996, 196). In recent years the Ford Foundation has worked with a network of Brazilian business leaders to stimulate a new tradition of corporate giving, but philanthropy remains the exception rather than the rule. In the survey conducted in 1993, almost 83 percent of all organizational funds were from international sources, with 76 percent of the total from international NGOs and the remainder from bilateral and multilateral agreements with foreign governments. Just 2.4 percent of ABONG's members' funds were from individual donations and 1.8 percent from businesses (Landim and Cotrim 1996, xviii). A survey in 1995 of the wider nonprofit sector (extended to include, among others, education and health agencies) found that 73.8 percent of the sector's funds were from fees and charges, 15.5 percent from the Brazilian government, and 10.7 percent from philanthropic sources (Landim, Beres, List, and Salamon 1999, 404).

As environmentalists began to create professional NGOs during the second half of the 1980s, they both reflected this larger picture and challenged it in other ways. The organizational changes provoked consternation among environmentalists who valued the ethical commitment that they believed was embodied in a small-scale, local, voluntary model of organization. Professionalization did make it possible to sustain positive, enabling action, to raise and administer funds for more ambitious projects than volunteers could carry out. But maintaining professional staff and organizational headquarters made NGOs necessarily reliant on some combination of governmental, corporate, and international financing. Environmental NGOs that made this choice were initially viewed with suspicion by small environmental organizations, but gradually were accepted as important clearinghouses for information, besides running their own campaigns and activities.

The third wave of activism was accompanied by considerable debate about this development. According to partisans of professionalization, at issue was replacing the ecology of protest (*ecologia de denúncia*), which aimed simply to block policies, with an ecology of results (*ecologia de resultados*), which aimed at enabling more positive accomplishments (interview with Capobianco 1990). Advocates of the ecology of protest regarded the stances taken by advocates of the ecology of results and the tactics used by them as excessively pragmatic; their own preferences were purist or authentic, in their view (interview with Sorrentino 1990). All the activists were pondering how to make the best use of newly opened channels of political access, whether the opportunities originated in domestic democratization, keystone events like the Rio Conference on Environment and Development, or newly available funding from businesses and the state.[3] We use examples of two organizations to illustrate how these choices could play out, and then present a brief overview of the organizational characteristics of the environmental sector in Brazil as it changed over time.

If the FBCN was the paradigmatic "old-style" conservation organization in Brazil, the SOS Atlantic Forest Foundation (SOS) was the most dynamic of the new ones formed in the late 1980s and early 1990s. The few existing environmental foundations had been set up either by university professors or by former government officials: FUNATURA, for example, was founded by Maria Teresa Padua after she left the IBDF. By contrast, SOS represented a shift in strategy on the part of a group of activists from existing ecological movement organizations. SOS sought autonomy, professionalism, and pragmatism in furthering environmental goals, all of which required institutionalization. Its early leaders included the second-wave activists Capobianco, founder and president of the Association for Defense of Juréia, and Feldmann, founder of OIKOS. They brought in Rodrigo Mesquita of the newspaper *Jornal da Tarde*, who was engaged in ecological struggles on the São Paulo coast, and the industrialist Roberto Klabín, a member of OIKOS. Capobianco argued that without technical personnel of its own, the Juréia association could never defend itself in confrontations with the government. Therefore SOS had to be "a professional institution that involves *técnicos* in maintaining a disinterested professional staff outside the government that could produce information and do long-term work on behalf of the Atlantic Forest. We agreed that it could not be made up only of environmentalists, and would have to be a pragmatic organization" (interview with Capobianco 1993). In 1993 the SOS administrative council had sixteen members, of whom, according to the bylaws, one-quarter were environmentalists, one-quarter scientists, one-quarter businesspeople, and one-quarter media people.

The SOS budget grew rapidly until the early 1990s, from $24,000 in the first year to $800,000 in 1992. The organization refused to take money from the state, but did accept money from private corporations. This confused other environmentalists, who suspected that corporate contributors controlled the organization's agenda. Objections to the corporate connections only began to recede during the preparatory process for the 1992 UNCED conference, when Capobianco was one of two coordinators of the Brazilian NGO Forum's secretariat (discussed below).

SOS was the only conservation foundation established in the 1980s that invested in building a membership base large by Brazilian standards, and by 1993 it had 6,600 dues-paying members.[4] Membership eventually allowed the organization to fare better than many of its counterparts as foreign donor funding became much harder to get. By 1995 SOS had more than ten thousand contributing associates, but eight thousand of these were associates by virtue of holding SOS affinity credit cards from the bank BRADESCO. These eight thousand signed up completely independently of the organization, which did not even have a list of their names. By then contributions from members and associates represented between 60 percent and 70 percent of the organization's revenues. Another 20 percent were accounted for by annual contributions of $2,000 from sixty individual businessmen and women. And 5 percent came from the sale of t-shirts and other items with the SOS logo, often produced by a corporate partner that might then put the corporation's logo on the item as well (Mantovani in Mafra ed. 1995, 99–100). According to the organization's web page (www.sosmatatlantica.org.br), in April 2006 membership stood at seventy thousand.

Many SOS founders believed that their Atlantic Forest campaign should—and could—be insulated from politics, by which they meant putting partisan or particular interests above effective public management. However, recognizing in 1988 that the organization had to become more "political" in the sense of winning allies and support for its activities, the board asked Capobianco to take over as superintendent, moving him from the council to the staff. An example of how this shift in perspectives affected organizational action was in the debate over debt-for-nature swaps. According to a pragmatic view, swaps were simply another way of capturing resources. But most other environmental organizations opposed debt-for-nature swaps, and the desire to maintain good relations with others argued for non-involvement, the path that Capobianco favored (and that the organization eventually took).⁵ The council also occasionally turned down offers of corporate money, fearing that some funding sources would cause the organization to lose dues-paying members. In 1991, when financial difficulties forced SOS to lay off a quarter of its twenty-five employees (responsible for policy questions) and one of its directors (responsible for technical aspects of project management), it kept the policy director.

The SOS position that government money came with too many strings attached was unusual among environmental organizations. When SOS turned down money from the government forestry institute, IBDF officials were shocked. During a campaign in 1991 for a cleanup of the Tietê River in São Paulo, which brought about a huge mobilization and perhaps the biggest petition drive in Brazilian history, government agencies signaled a number of times their desire to participate, and were refused. Capobianco explained, "Independence is really important. When you get money from several firms, what can a firm do? At worst, they can say they won't give you any more money. OK, fine. But when you get money from the government, the conflict with the government is fierce. A conflict with a firm is about something very specific, like pollution or deforestation, so it's very clear what's at stake. With the government it's harder. . . . How can you criticize the government for omission if you are funded by the government—it confuses people . . . So we don't take government funding, and international financing is hard to get" (interview with Capobianco 1993).

SOS received support from such donors in the United States as the World Wildlife Fund, Conservation International, the John D. and Catherine T. MacArthur Foundation, and the Ford Foundation. But foreign donors are fickle and their agendas change. When SOS began, its American donors, for example, did not support projects aimed at local communities—they wanted parks and reserves. Then suddenly the foundations' priorities changed, and everything became community-focused. The speed of the shifts in priorities spoke for a diversified funding base, but there were few options for the first professionalizers. From 1991 to 1994 SOS established a partnership with Radio Eldorado and Unibanco for a major campaign to clean up the Tietê River in São Paulo, coordinated by a Núcleo Pro-Tietê. Capobianco left SOS after the 1992 Earth Summit to become a founding director of ISA, the Instituto Socio-Ambiental. In 2003 he became secretary for biodiversity in the Ministry of the

Environment. Although SOS continued to agitate for cleanup of the river basin, the departure in 2001 of its main campaigner, Samuel Baretto, for WWF pulled it into broader collaboration in this area, reminding us of the important role that individual trajectories have in configuring inter-organizational projects and relations.

SOS Mata Atlântica remains one of the largest organizations of its kind in Brazil. It has a corps of professionals working on environmental education, monitoring of forests and bodies of water through satellite imaging and direct observation, protecting native species, maintaining information databases, and so on. Its work is in part direct environmental practice, in part educational, and in part the mobilizing of many volunteers to participate in tree planting and monitoring activities. The organization also belongs to seven forums and councils, ranging from ABONG to the Forest Stewardship Council.

In contrast to SOS Mata Atlântica, the pioneer environmental group AGAPAN opposed professionalization based on private funding; its refusal to take money from "polluters" eliminated most potential corporate donors. Instead AGAPAN turned more to the political side for financial support, and illustrated some of the dilemmas to which Capobianco referred. In 1990 two PT municipal councilors, Giovanni Gregol and Gert Schinke, gave AGAPAN 10 percent of their salaries, thus providing most of its budget. AGAPAN's president Celso Marques justified this on the grounds that AGAPAN's political force had won the election for Gregol and Schinke, and that as AGAPAN members they would take its views to the municipal council (interview with C. Marques 1990).

If the "tithing" of Gregol and Schinke made some identify the group with the PT, AGAPAN's founder José Lutzenberger confounded everyone by accepting the unexpected invitation from President Collor to be secretary of the environment. Running as a right-wing modernizer with a strong dose of populism befitting a "national savior," Collor beat out the PT leader Luís Inácio Lula da Silva in a close runoff election in November 1989. Participants in a statewide gathering of environmental organizations in September 1990 in Santa Maria, Rio Grande do Sul, were so polarized over Lutzenberger's role in the Collor government that every topic discussed all weekend turned into a debate over whether the movement gained strength and influence by participating in the government or remaining independent from it.

In fact, although some members of AGAPAN became politically prominent, the organization never raised enough money to play the roles that NGOS with more resources could play. In 1990 AGAPAN had only one part-time employee. It could not send representatives to all the myriad gatherings of full-time environmentalists, nor could it propose and implement large projects (interview with C. Marques 1990). Like other smaller volunteer organizations, it continues to work locally and at the state level, while more professional organizations occupy positions of national visibility. After the death of José Lutzenberger in April 2002 its presence further diminished, as new, young leadership had not taken the place of their distinguished predecessors.

After two decades of professionalization, the typical organizational profile is closer to AGAPAN's story than that of SOS: a majority of Brazilian environmental organizations was either unable or unwilling to professionalize, notwithstanding the impression even among activists that the volunteer period was over (Svirsky and Capobianco eds. 1997). A survey of such organizations in 1994 found that 70 percent of them had had no activities requiring paid employees, even occasional ones; of the seventy-two organizations that had some paid employees, forty-three had fewer than ten and only three had between twenty-six and thirty-five (Pizzi 1995, 43).[6] Figures from 2002 support this picture (see table 3.1). Nearly sixteen hundred environmental organizations had an average of just 1.61 salaried employees each, meaning that many must have had none at all, although the figures are not disaggregated. Environmental organizations lag behind other NGOS in their employment figures except for NGOS in the Amazon region, where there are nearly six employees per organization. Most of these are in the environmentally contested states of Rondônia and Pará. Given these employment figures, it is not surprising that 43 percent of the respondents to the earlier survey should have claimed a budget of under $10,000 for 1994 and that only five of 216 organizations in the survey had budgets of over half a million dollars (Pizzi 1995, 41).

Many environmental groups opted out of professionalization, recognizing that it has costs as well as benefits. Donors rarely want to contribute to organizations per se, but rather to projects that the organizations will carry out. Some even stipulate that none of their contributions be used to pay for organizational overhead, hoping thus to maximize their impact on a particu-

Table 3.1 Employment in NGOs for the Protection of the Environment and Animals, 2002

	# OF LOCAL ORGS	# OF SALARIED EMPLOYEES	AVG. # OF EMPLOYEES PER ORG.
North (Amazon): All NGOS	11,715	46,484	3.97
North: Environment NGOS	101	604	5.98
Northeast: All NGOS	61,295	215,371	3.51
Northeast: Environment NGOS	190	418	2.20
Southeast: All NGOS	121,175	857,633	7.08
Southeast: Environment NGOS	809	1,219	1.51
South: All NGOS	63,562	306,367	4.82
South: Environment NGOS	365	294	0.81
Center-West: All NGOS	18,148	115,435	6.36
Center-West: Environment NGOS	128	29	0.23
Total, Environment NGOS	1,593	2,564	1.61

Source: Calculated from IBGE 2002, table 8.

lar project's activities. Yet without institutional funding or endowments, professional organizations that want to hold on to skilled staff and infrastructure have little choice but to seek project funding that will allow them to pay their own costs—a new computer here, reams of paper there, a jeep for monitoring activities somewhere else—and salaries must typically be moved among projects as well. This can promote a tendency among professional organizations to shape their agendas according to the kinds of projects that are being funded, a problem that some of Brazil's most prominent environmental organizations lamented at a meeting in 1995 called by the Damian Foundation (Mafra ed. 1995).

The exact funding profile of those Brazilian environmental organizations that have pursued funding has several unexpected dimensions. In the broadest study of nonprofit organizations already cited above, the sector comprising environmental and other (undifferentiated) advocacy groups stood out as the only one for which philanthropy was the dominant source of funding, at 73 percent of total revenue, with the remaining 27 percent from the public sector and none from fees (Landim, Beres, List, and Salamon 1999, 407).[7]

This unexpected result is supported by the survey conducted in 1994, which found that members' contributions were the largest source of funding for just over half the organizations. Almost as many had received funding from businesses as had received international funding (Pizzi 1995, 41), although it should be noted that this figure measures only the number of organizations and not the total value of contributions.

Three-quarters of the NGOs analyzed received no money from international sources. Those who did receive international funding were most likely to work on agro-ecological or socio-environmental matters, with conservation organizations close behind (Pizzi 1995). The larger environmental and indigenous support organizations (three-quarters with annual budgets over $50,000) who were members of ABONG averaged almost 60 percent of their funding from international NGOs in 1994 (calculated from Landim and Cotrim 1996). None of these are traditional conservation organizations. These results seem to confound both the understandings that many Brazilian environmentalists have about international funding practices and more general studies such as that by Lewis (2003), which stress that international donors emphasize global environmental concerns such as biodiversity rather than local concerns. The number of organizations that received international funding for Amazon conservation, especially through the G-7 Amazon project, undoubtedly increased after 1994. In any case, the data suggest that professionalization transformed a small but especially visible and active set of groups that tend to fit either the "new conservation" model of SOS or the socio-environmental model of FASE, IBASE, and the Amazon's rubber tapper organizations, which include the environment as just one of a set of social concerns.[8] For them and groups that aspire to be like them, international funding is critical and presumably shaped by donors' priorities—but those are more complex than often thought.

The subset of professionalized NGOs remains a favorite target for ultranationalists, who regularly accuse them of being instruments of foreign governments. There are also complaints that NGOs' growing skill at getting the attention of media and influential actors leads them to take on roles that representative organs instead should play. Amazonian politicians (seeking to evade criticism for excessive deforestation and illegal timbering) initiated congressional hearings on oversight of NGOs that ran from 2001 to 2002, giving rise to yet another draft bill regulating their activities and mandating reporting requirements that essentially duplicated those already in force.⁹

Socio-environmentalism

Although AGAPAN resisted one of the innovations of the third wave of Brazilian environmentalism, it embraced another—the development of a discourse and, more slowly, a practice of socio-environmentalism. By the beginning of the 1990s most Brazilian environmental organizations had adopted a social environmental discourse, one that made poverty and environmental degradation part of the same causal story (Stone 1989). In our interviews with leaders of environmental organizations from Rio de Janeiro to Rio Grande do Sul in 1990 and 1991, almost all spoke of the need to link the environmental and the social—whether or not they could point to examples in their own work. What that meant concretely was not always clear. In some cases it meant that environmental degradation affecting the poor should receive the most attention, in others that the poor should be "brought into" environmental work. In its strongest sense, it meant a belief that in the realm of development favoring the poor and protecting the environment were mutually reinforcing goals.¹⁰ Still, although most urban environmentalists had no trouble linking environmental protection to livelihood struggles in the Amazon—of rubber tappers, artisanal fishermen and women, Brazil nut gatherers, and the like—they initially found it much harder to integrate environmental struggles with poor peoples' movements in cities. The language of environmental justice, perhaps the most successful frame of this kind, was not widely used in Brazil until the next decade (see chapter 5).

Three explanations of why socio-environmentalism appeared when it did merit further reflection. The new environmentalism emerged as the military dictatorship was ending, with an attendant rise in organizing initiatives within civil society in general, especially on the left. Second, the murder of Chico Mendes in Acre in 1988 generated widespread discussion of the links between the livelihood struggles of traditional forest peoples and protection of the Amazon. Third, the preparatory process for the United Nations Conference on Environment and Development (or "Earth Summit") in Rio de

Janeiro in 1992 brought a wide range of environmental organizations, women's organizations, urban and rural trade unions, and other social movements together in sustained dialogue over almost two years.

Democratic Transition

We have already seen that the Brazilian environmental movement was self-consciously political (though not necessarily partisan). In the Brazil of the democratic transition, turning one's back on politics implied accepting an oppressive status quo. This differentiated the Brazilian movement from many of its counterparts in the northern hemisphere. For his book *Como Fazer Movimento Ecológico e Defender a Natureza e as Liberdades* (1985), Carlos Minc sent questionnaires to 650 ecological groups throughout Brazil, and received 200 responses. On the basis of these he concluded that "People describe the so-called ecological question quite differently in our country and in the developed world. Here the ecological question is intertwined with the drama of exploitation and misery that rich countries in the main do not have" (Minc 1985, 65). Most environmentalists believed that they could neither solve Brazil's environmental problems nor develop workable political coalitions in Brazil's new democracy without confronting social issues.

Political contexts mattered. In the industrialized world, a significant sector of the environmental movement was explicitly engaged in a critique of the traditional left—in Western Europe a critique of social democratic and communist parties, and in the United States a critique of aspects of the new left student and antiwar movements of the 1960s. Although we might classify some of these as left-libertarian (Kitschelt 1993), they shared a critical stance toward the Marxist left, one that became even more pronounced in former Soviet bloc countries. But in Brazil the new environmentalism and the return of the left to the political stage occurred simultaneously, and were frequently intertwined.

For many environmentalists, the Workers' Party became a political home not so much because of its socialism (though many environmentalists were also socialists) but because it espoused radical democracy, direct participation, and grassroots organization (Keck 1992). Although the PT generally treated environmental protection as one of a laundry list of social demands, it was still the only party in which programmatic questions were debated seriously by party members, who could vote to change party positions. (On PT environmental debates see Benjamin 1990; Benjamin 1991; Buaiz 1988; Minc 1991; Viana, Silva, and Diniz eds. 2001; Waldman 1990). In the late 1980s and early 1990s ecologists taking part in electoral politics vacillated between the Green Party and the PT (interviews with Waldman 1990; Vieira 1991). Even when many ecologists left the PT in disgust, either at anti-ecological positions or at sectarian behavior, they did not reject the principle that ecological questions had to be linked to social justice.

Chico Mendes

A second major factor in the development of social environmentalism was the death of Chico Mendes (discussed more fully in chapter 4). Support for Amazonian rubber tappers was widespread among Brazilian environmentalists well before 1988. By the mid-1980s the CNDDA, based in Rio de Janeiro, had established committees for defense of the Amazon in almost all states and major cities of Brazil. It developed a relationship with Mendes's Acre group early in the group's history. So did the organizations in which PT members predominated, as the rubber tappers, including Mendes, played an important role in organizing the party in Acre and were active in the rural labor movement (Keck 1995). Still, before 1988 most people saw the rubber tappers' struggle as a labor and land struggle, and only secondarily saw it through an environmentalist lens. After 1988, that changed.

The shift occurred partly because of the alliance between the rubber tappers and environmentalists in Washington who were working to influence multilateral bank lending (Keck 1995; Fox and Brown 1998; Bramble and Porter 1992). But other things going on that year increased the importance of these links. Here is an instance where *simultaneity*—or coincidence in time—clearly plays an important political role. Internationally, 1988 was the year when the Amazon burned. The unusually hot summer and drought in politically important parts of the United States combined with satellite data on the extent of fires in the Amazon to produce media images of the forest going up in flames. The press and many environmentalists portrayed Brazil as an environmental rogue, and deforestation as a significant contributor to global warming. Chico Mendes's murder two days before Christmas fed into this already overheated situation and gave the debate over deforestation a human face. It also gave environmentalists an identifiable hero—someone whose "fight to save the forest" (Green ed. 1989) broached social and environmental questions at once. Keck has discussed at length elsewhere the international side of this campaign (Keck 1995; Keck and Sikkink 1998a).

Mendes's murder also weakened some of the Brazilian government's claims about purported foreign efforts to encroach upon Brazilian sovereignty in Amazonia (Hurrell 1992). President Sarney's use of the nationalist card in the late 1980s would normally have had support on the left. As Warren Dean put it in his usual trenchant manner, "Why, the nationalist left has asked, has the preservation of Amazon forest achieved such instant resonance among a foreign press that exhibits so little concern for the forests and indigenes of its own country" (Dean 1989, 3). Since Mendes and his group were undeniably part of the left, the position described by Dean was an increasingly untenable one.[11]

The United Nations Conference on Environment and Development

The third factor that fomented the development of socio-environmentalist positions was the same one that brought them most relentlessly into engagement with quite differently positioned foreign organizations, institutions, and ideas: the two-year preparatory process among Brazilian organizations leading up to the United Nations Conference on Environment and Development in Rio de Janeiro in 1992 (known variously as UNCED, the Earth Summit, and Eco 92). This was a multilevel process, entailing shifts in a whole series of relationships among Brazilian environmental organizations, the Brazilian government, international institutions, foreign governments, and foreign environmental organizations.

The Earth Summit, Rio de Janeiro, 1992: Brazilian Foreign Policy and the Environment: From Stockholm to Rio

During the 1970s the Brazilian government held to the position it had taken at the UN Conference on the Human Environment at Stockholm in 1972: that environmental questions should not affect other issues in international relations, that they fell within the sovereign jurisdiction of each country, and that they were of secondary importance for developing countries (Campbell 1973; Guimarães 1991). This position began to change mainly as a result of pressure from social movements and NGOs—both Brazilian and foreign— and from the governments of developed countries. In 1987 and 1988 Brazil was subject to intense pressure to control the rate of deforestation in the Amazon. Brazil's most common response was either to assert that the pressure threatened Brazil's sovereignty, or to argue that as long as the industrialized powers were doing most of the polluting, they were not in a position to criticize Brazil (interview with Fonseca 1990). Just as the Médici government had done at Stockholm in 1972, the Sarney government hoped to portray international reaction to the Amazon fires as an attempt by foreign powers to "internationalize" the Amazon, to craft a shield behind which foreigners could hide their designs on the region's wealth (Hurrell 1992; Kolk 1998).

Certainly many foreigners were markedly insensitive to language as they took up the issue, disparaging not only Brazil's political will but even its fitness to regulate affairs within its own territory. At a conference at The Hague in March 1989 that debated the desirability of a supranational authority with sanctioning power, President François Mitterrand of France pushed all the buttons: "This will result in a loss of sovereignty for some countries, but it has to be done." Although he went on to say that European countries had already agreed to submerge some of their sovereignty within the European Community, what Brazilians remember is newspaper headlines like "Mitterrand Wants Countries to Renounce Their Sovereignty" (*Jornal do Brasil*, 12 March 1989). But Brazil's seeming defiance of the Euro-American Goliath seemed hollow after Chico Mendes's death, when a more credible David captured the moral high ground of the debate.

At the end of 1988, when the UN General Assembly passed its resolution convoking a UN Conference on Environment and Development to be held in June 1992, Brazil, Canada, and Sweden announced their candidacies to host it. President Sarney made the decision himself, at the suggestion of Itamaraty (the ministry of foreign relations), hoping by its bid for the conference to demonstrate a serious desire to remedy its environmental problems (interview with Bitelli 1990; Kolk 1998, 1485). By the time Fernando Collor de Mello took office in 1990 as Brazil's first directly elected president since 1960, it had already been decided that Brazil would host the international meeting.

The Collor government took a more assertive stance on the environment than its predecessor. Collor considered himself in a position to disclaim responsibility for the results of past governments' actions, saying that he would turn a new page in environmental policy (interview with Fonseca 1990). He sent a strong signal to this effect with his appointment for environmental secretary. Apparently during Collor's travels abroad before taking office, in conversations about the environment everyone from Prince Charles of Great Britain on down mentioned José Lutzenberger of AGAPAN (interview with Hathaway 1990). Upon his return to Brazil, Collor told Lutzenberger that if he would agree to become environmental secretary, he would have a lot of room to influence policy. Collor went on to make Brazil's actions on the environment a central part of his diplomatic strategy (Fonseca 1990; Bitelli 1990). He created high-profile photo opportunities, such as one at an illegal airstrip in Yanomami territory that was ostensibly being demolished at his behest. This particular effort at public relations failed, thanks to modern communications. By the time Collor came to Washington soon after the photos had circulated, environmentalists there had received e-mails and faxes from Brazil telling them that the airstrip had already been quietly rebuilt. President George H. W. Bush had this information by the time he met with Collor (interview with Bramble 1990). Still, the appointment of Lutzenberger and other signs of attention to the environment gave the Collor government, for a time, a veneer of environmental respectability.

The Brazilian Government in the Organization of the Earth Summit

The Brazilian government treated the Earth Summit more as a diplomatic than as an environmental or development gathering. Itamaraty dominated the preparations and many of the governmental positions. But it was unable to establish a single voice with Lutzenberger, at the helm of SEMA, who was drawing extensive international and national press attention for his radical environmental views. In a memorable briefing for NGOs at the 4th Preparatory Meeting in New York in March 1992, for example, he railed against corruption in IBAMA (under his official control) and told international NGOs not to give any money to Brazil because nefarious practices would make it disappear.[12] After the newspaper *Folha de São Paulo* reported on the meeting—videotaped by an attending environmentalist—Lutzenberger was removed for the phlegmatic official reason that it "must be assured that there is just one thought and action in the area of environmental protection in the government" (cited in *Folha de São Paulo*, 22 March 1992). With Lutzenberger gone, the domination of the Brazilian negotiating positions by career diplomats was virtually complete. These career diplomats were by tradition and training disinclined to seek national citizens' input into their negotiating positions or to promote strong environmental stances. The government started its required analysis of national environmental conditions quite late, bringing Roberto Guimarães back from CEPAL in Santiago to coordinate the process, in which "experts" on various topics were invited to contribute portions of text. Itamaraty was reluctant to invite public debate, even in the legislative branch.

The Brazilian NGO Forum

Brazilian environmentalists were disappointed with their government's approach to the conference.[13] Nonetheless, they were very conscious of their role as hosts of an international gathering of NGOs. They organized their own participation in the Brazilian NGO Forum in Preparation for the Civil Society Conference on Environment and Development, which eventually included around twelve hundred organizations from many sectors of civil society (Fórum de ONGs Brasileiras 1992). They sought and were given representation in virtually all international NGO preparatory forums, with their participation financed by UN-related organizations, international NGOs, and even foreign governments. Thus Brazilian environmentalists were able to travel the globe to attend preparatory events, without needing to rely on and relate to the Brazilian national delegation. International NGOs also flooded Brazil in the year before the conference, establishing temporary offices and providing additional informal contacts between Brazilian and northern NGOs. In 1992 all began intense planning to organize a massive, parallel nongovernmental conference, eventually known as the Global Forum.

Aside from traditional forms of exchange and debate, the preparations for the Earth Summit were strongly marked by the rapid spread of communications technologies. When the preparatory process began, e-mail was in its infancy and the World Wide Web did not yet have a public interface. Exchanges of documents took place by fax if they were urgent, and by slower forms if not. Environmentalists communicated through listservs and bitnet newsgroups. Because of Carlos Afonso's early work on computer technology at IBASE, Brazil became the first country in Latin America and one of the first in the developing world to have a relatively low-priced subscription service to the Internet, called Alternex. The network provided e-mail and access to newsgroups, both its own and those of the Institute for Global Communications, which developed gateways to other countries. Most early subscribers in Brazil were organizations rather than individuals, but by the beginning of the process of preparing for UNCED, it was possible to post news to a newsgroup in one part of the world and have it read everywhere. Thus from the very beginning, both the official documents from the conference and the nongovernmental debates were available. Moreover, a U.S. citizen then living in Brazil, Langston James Goree III (known as Kimo), who had been sending extensive descriptions of the first Prepcom in Nairobi by e-mail to his friends, who sent them on to many others, then started a news service, Earth Negotiations Bulletin, that has been covering UN environmental conferences ever since.[14] The intensity of communication only increased over the two years leading up to the conference, as Internet access became easier and modem connections faster.

As Brazilian NGOs discovered at least some northern NGOs willing to discuss questions of global equity that the governments were determined to avoid, they began to see more transformative potential in a global coalition of citizens than in the official conference. Participants in the Brazilian NGO Forum became vital adherents of a belief that the United Nations conference was an opportunity for articulations among citizens rather than with governments (Friedman, Hochstetler, and Clark 2005, 39–40). They wanted to use the Global Forum to begin popular mobilization on a global scale. They believed that otherwise any documents produced in the conference would remain empty pieces of international legislation, like their own constitutional chapter on the environment.

According to Tony Gross, a founder of the forum and one of two initial executive secretaries, the Brazilian NGO Forum began to get organized when he and leaders of OIKOS and SOS-Mata Atlântica realized that if Brazilian NGOs did not define their role in the events, others would define it for them (interview with Gross 1990). Gross worked with CEDI, a large social development NGO founded by Protestant churches. A small group met in March 1990, and then issued a letter in May inviting other organizations to a larger meeting. In their letter the three instigating NGOs argued that since the Brazilians would host the parallel conference, they had to set the terms of debate on major issues, and were responsible for keeping the predictable North-South divisions from stalemating the discussion. They wanted the forum to be able to coordinate preparations by NGOs for conference discussions and the logistics of the parallel meeting (CEDI, OIKOS, SOS 1990). If Brazilian organizations were to claim the right to name their own representatives to international conclaves, instead of waiting for invitations from organizers abroad, they had to organize fast. Further, they wanted to prevent a contending group which became known as the Pro-Rio coalition from occupying the space that Brazilian NGOs should occupy themselves.

Gross managed to "push his way in" to the next meeting called by the Center for our Common Future at Nyon, Switzerland, in June, with representatives of eighty-six organizations from thirty-three countries. From that meeting came a decision to establish a multi-sectoral International Facilitating Committee to focus "independent sector" activities at Eco '92, on which Gross secured a place as provisional representative of the still-incipient Brazilian NGO Forum (interview with Gross 1990; IGC posting in newsgroup en.unced.general, 8 September 1990, by rpollard). During the first stages of the organizing process, Gross was probably the best-prepared of the Brazilian activists to play this kind of political entrepreneur role in relation to much larger and better-organized foreign NGOs. However, his bridging role went beyond linking domestic and international: he also spoke the languages of both environmentalists and social development organizations. Gross arrived in Brazil in the 1970s from Great Britain, initially to do fieldwork for a doctoral dissertation on land conflicts in Acre. Witnessing the conflict at its height, he became actively engaged in defending the rubber tappers' movement, and went to work for Oxfam. In 1989 he joined CEDI, known for its work in support of indigenous populations and other popular movements. Active integration of environmental and livelihood concerns had become central to rubber tappers' organizations in the mid-1980s, giving Gross plenty of practical experience in making those links. Rubens Born from OIKOS, a long-time environmental activist in São Paulo, became the Brazilian NGO Forum's designated representative in future meetings for similar reasons, both because of his history and because of his possession of the sine qua non at those meetings, fluency in English.

In the meantime, Brazilian nonstate actors struggled over who would represent them and host the nongovernmental side of the UN conference. The original General Assembly resolution had called for broad participation by NGOs in preparations for the conference, but delegates to the first preparatory conference in Nairobi (6–31 August 1990) had not even agreed on a credentialing process. Maurice Strong, the conference's secretary general, joined NGOs in promoting an inclusive policy, and most of the NGOs that managed to get to Nairobi were able to participate. In March 1990 Strong met in Vancouver with representatives of NGOs and the Brazilian government to discuss the NGOs' participation. Brazilian organizations present included Salve a Amazônia (Save the Amazon), a fairly new organization, which, according to Gross had simply heard about the meeting and turned up by paying its own way. Salve a Amazonia was subsequently an instigator of a "Pro-Rio" coalition of businesses, some entrepreneurial environmentalists, municipal authorities, and nonprofits in Rio that set out to take over preparations for the event. The group mobilized businesses to provide resources for the event, and formed a council of local notables to mobilize other sectors of society. Pro-Rio contacted the United Nations and with the Center for Our Common Future began to gather information from other organizations regarding plans for the meeting (interview with Brito 1990).[15]

Many Rio environmentalists were outraged. A meeting of APEDEMA in Rio de Janeiro on 20 September 1990 produced a storm of condemnation of Pro-Rio's claim to have broad support from environmental organizations, whose involvement seemed to be precluded by Pro-Rio's bylaws. Brito stood up to refute some of these accusations, saying that mistakes had been made and were being corrected, but also stressing that Pro-Rio had never been intended as a democratic organization but instead as an operational one— the role that in part it eventually played. Others argued that Pro-Rio should be explicitly deauthorized, noting that at the first meeting of the governmental preparatory commission in Nairobi, Rubens Born had a hard time convincing people that he was there as a representative of Brazilian NGOs, as Pro-Rio had already occupied that space. APEDEMA during this period was being revitalized by the active engagement of some young Turks of the Rio environmental movement. Activists like Rogério Rocco, of Os Verdes, and Franklin Matos, of Grüde, were part of a cohort of young activists, who were associated with organizations that were still getting off the ground, but who were (and remain to this day) thoroughly dedicated to the environmental movement.[16]

Meanwhile, the organization of the forum had moved quickly and reached more broadly than the initial organizers had expected. The first meeting to discuss coordination of the NGOs, held in São Paulo at the end of March 1990, included forty invited groups. It was followed by a second meeting at Novo Friburgo on 30–31 July with representatives of sixty groups. From then on the numbers snowballed as each organization contacted others. Although the organizers expected that other kinds of civil society groups in Brazil would establish forums of their own, the Brazilian NGO Forum became an omnibus organization of environmentalists and a variety of other social action and labor groups, ranging from the Unified Workers' Confederation (CUT), to feminist, social action, and human rights groups of varying size and scope, to professional NGOs engaged in "action research." The forum's one restriction on membership was to define NGOs so as to exclude business groups, in contrast to the United Nations. It did so by requiring that participants be "recognizably independent in relation to the current model of development" (Fórum de ONGs Brasileiras 1990, 1). Each organization had one vote in the forum, whatever its size and whatever its capacity to undertake responsibilities outside the meetings. For some of the larger professionalized organizations this was irritating; on the other hand, smaller environmental organizations viewed the big professionalized organizations as "King ONGs" (ONG being the Portuguese acronym for NGO) trying to crowd out environmental activist groups. These changes of the third wave were still in full and contentious development. Similarly, it took quite a while for some organizations in the forum to develop good relations with mass organizations like the CUT.

The coordinating committee and secretariat (composed of CEDI and UPAN, from Rio Grande do Sul) faced enormous logistical problems. These included not only a lack of money with which to organize national preparations and accompany international ones but also a lack of accessible information. For example, UN documents for the conference had not yet been translated into Spanish, much less Portuguese, and thus were not yet accessible to most Brazilian organizations. Decisions had to be made about which documents the forum wanted to produce, and how to relate them to the production of Brazil's National Report, being coordinated by the secretariat of the environment through Itamaraty. Further, they had to decide how to relate to coalitions like Pro-Rio, which remained important in logistical organization of the city.

Early meetings of the forum were mainly concerned with procedural questions, like who should participate and how decisions should be made. Some ecological groups (for example the coordinating committee from the Northeast) objected to the role of omnibus social action groups like CEDI. Worry about the role of the "King ONGS" grew stronger when IBASE joined the forum, the coordinating committee, and the executive secretariat in June 1991. This conflict over coordination came to a head at the fifth national meeting in Rio, where IBASE's professional style conflicted with the more youthful voluntarism of the APEDEMAS from Rio and São Paulo. Environmentalists met separately in September of that year to decide if the forum process merited their continued involvement, before concluding that their own stronger participation would improve its results.

Here the long-standing identity politics of the environmental movement activists contrasted with the pragmatic, means-ends logic that the professional organizations used to match tasks with capabilities. Although the process of creating professional organizations had begun before UNCED, it accelerated afterwards; nonetheless, the militants' traditional mistrust of professionalism remained very important during the early stages of the forum. Frustrated with what he saw as fruitless "struggles over power and hegemony," the executive director of IBASE, Herbert de Souza (Betinho), submitted a letter to the forum announcing his resignation from the national coordinating committee and the secretariat, a letter whose contents were more than a little condescending regarding the "sterile and paralyzing opposition" to IBASE among smaller organizations. Its disparaging tone notwithstanding, Betinho's letter pinpointed a central difficulty of the forum's continuing effort to develop common positions. He wrote:

> The Forum's diverse makeup can be either its strength or its weakness, depending on how this heterogeneity is expressed within a context of transparency and democracy. This is the greatest challenge whenever dealing with far-reaching projects involving diverse social actors. The Forum's makeup includes highly differentiated fields of environmental organizations, NGOS and social movements. In spite of its diverse composition, the Forum has not succeeded in consolidating the social and political scope that diversity could make possible. False dichotomies have been championed, such as the one separating "environmental organizations" and "Social Development NGOS." These have made no contribution to substantive progress in civil society's discussion of the subject at hand, namely "environment and development." (Souza 1991)

Ultimately most participants did decide to stay in the forum, and it became the accepted organizing structure for participation by Brazilian NGOS in the UN conference. As such, it was also the home for broad substantive debates as well as debates over organizational concerns.

The Substantive Debates

Gross and Born had became convinced in Nairobi, while attending the first official Preparatory Conference (Prepcom) for the UN Conference in August 1990, that the Brazilians had to start preparing and circulating position papers on substantive debates if they were to have any chance of influencing the process. They hoped that they would begin to do so before the second Geneva Prepcom in March 1991, and they expected that producing and debating position papers would help to identify which organizations had the capacity for this kind of work. They had received $80,000 in Canadian funds for the coordinating commission, and hoped to find more to cover the cost of writing and distributing position papers and holding meetings of the forum.

In early debates, most conflicts stemmed from the differences in worldview between ecological organizations on the one hand and NGOS like CEDI and FASE on the other, which wanted to focus on the connections between environment and development. The organic linkages that were easy to see for the Amazon—the identity between environmentalism and the livelihood struggles of rubber tappers and small-scale fishermen and women, for example—were much harder to see in urban areas. Brazilian workers, like their counterparts elsewhere, feared that environmental regulation would endanger their jobs. Although Brazilian environmental organizations by then usually espoused a "socio-environmental" frame for their work, most had difficulty translating it into practice, except in terms of environmental education programs. Organizations without a tradition of environmental activism also had difficulty assimilating environmental ideas, as when the CGT labor confederation took the government line on the threat that the Amazon would be internationalized at the sixth national meeting in São Paulo in September 1991.[17] As long as there was little genuine dialogue between organizations working for the material and social improvement of living standards and those espousing an environmentalism of the poor, it was easy to maintain rhetorical unity by ignoring contradictory priorities. When the preparatory process for the Rio conference forced the two factions to forge common positions, the differences became much harder to ignore.

External events frequently produced deadlines for these discussions. Efforts to influence the agenda had to produce results by February 1991, for example, because at a meeting in Mexico City in March the G-7 countries were expected to take a position on the UNCED agenda. To advance a paradigm with sustainable development that was ecologically balanced, was socially just, and would eliminate economic distances between North and South, the forum needed to have a strategy to "make this the hegemonic position." They wanted to involve other social sectors, and to bring to Brazil a large number of participants in regional and global networks to take part in the forum. To propose a complex engagement of a very broad range of groups in a paradigm change suggested an extended process; the organizational timetable of UNCED made it something akin to a marathon run at a sprinter's pace.

National meetings were structured to foment national discussions on topics raised at the conference. There were two sets of concerns: accomplishing the goals that the forum had set for itself, and deepening participants' understandings of problems that they wanted to solve. By the sixth national meeting of the forum, held in São Paulo on 27–29 September 1991, they had formed working groups to discuss different aspects of the process. Six working groups were given responsibility for the forum's activities: organizing and supporting regional groups; producing technical background documents; drafting the alternative national report (subsequently renamed the civil society report); tracking the progress of the official national report; monitoring international conventions; and organizing the NGO conference. The report of this national meeting, and of other meetings, give the clear impression that the working group on organizing the NGO conference spent much time listening to the report of the national coordinators (after which most people left) and discussing measures to guarantee a representative sample of Brazilian NGOs at the conference, whether they could afford to pay for it themselves or not.

Although forum participants intended to prepare themselves to influence the international NGO agenda, their thematic debates focused on national problems rather than either the official conference agenda or how to construct an international NGO agenda. There were working groups organized around agriculture and the environment; sustainable development; models of energy use; population; environmental law; the semi-arid region; the urban question; aquatic ecosystems; Pantanal; savannah; the Atlantic Forest; Amazonia; the indigenous question; environmental education; environment and spirituality; Ilha do Bananal; and the concept of the NGO (for the final reports see Fórum de ONGs Brasileiras 1992). Most striking is the degree to which most (though not all) Brazilian environmentalists were much more intent upon discussing Brazil in the global system than the global system per se.

Brazilians Abroad

At the beginning of June 1990 the Center for our Common Future (Switzerland) formed an International Facilitating Committee (IFC) to facilitate participation by different "independent sectors" in the UNCED process. In its first two meetings the IFC expanded to include around twenty-five members, of which the Brazilian NGO Forum for UNCED was one (represented by Rubens Born). Although uncomfortable that the IFC's conception of nongovernmental organizations (called "independent sectors") included even private businesses with a history of polluting, the forum recognized that it was the only serious international effort to organize the parallel events at UNCED, and that only by participating could the Brazilian Forum avoid being crowded out and keep track of what the IFC was doing (Born in Annais VI).

The IFC members were worried too. While forum members were worried about keeping any one group from controlling the meeting, some of those at an IFC meeting in March 1991 in Geneva, like the veteran feminist leader Bella Abzug, worried that tens of thousands of people would have nowhere to go. The IFC president, Ashok Khosla of India, was quoted in the Brazilian newspaper *Globo* as saying, "If the Brazilians haven't realized that this is an international event, that it's not a Brazilian conference, we are going to have problems. They need to have a global vision. We are not interested in coordinating things. If they don't want our help, fine. But if thousands of organizations arrive in Rio and everything is not well organized, they are going to be furious" (*Globo*, 30 March 1991, 17).

With some misgivings, the forum accepted the recommendations from a meeting of national and foreign organizations in Rio at the end of May 1991 that the Brazilian NGO Forum work with the IFC to set up the Global Forum. In August the two established an international working group to plan the parallel conference, including NGO representatives coming from different regions (Africa, Asia, Europe, North America, Latin America), members of some international NGO networks, and representatives of the forum. The parallel organizing effort by the forum and the IFC had three goals: to lobby the official government conference in furtherance of proposals from civil society; to discuss an NGO agenda, with recommendations, proposals, agreements, and treaties; and to plan the future of NGO roles in the United Nations and regional organizations. Since the forum had only joined the IFC to guarantee this collaboration, the forum decided to withdraw from the IFC and concentrate on the activities of the International Working Group. Foreign NGOs found the attitudes of the Brazilians baffling. They worried about the logistics of the event, as they had gained the (correct) impression that the functioning of the NGO events was far from guaranteed. Many found the Brazilians' approach to substantive matters overly politicized, a perception that reflected discomfort with the leftist orientation of Brazilian NGO positions, while many of the large international NGOs involved in the process, like the World Wildlife Fund and Greenpeace, either called themselves nonpartisan or actively rejected the left's materialism.

The international working group created in Geneva in August 1991 during the third PrepCom was conceived as an operational task force, not as an organ of political coordination. Its only function was to organize the international NGO meeting to take place alongside the official UNCED conference in June. Nonetheless, many Brazilian NGOs expected the working group to provide political leadership and were frustrated by its refusal to do so, just as others worried about political manipulation and did not want to support any organization that might eventually try to speak in their name.

It is impossible to do justice to the story of the logistical preparations for the NGO conference. The organizers were frustrated at every turn—by the failure of the Brazilian government to provide support that most had assumed would be forthcoming, by the failure of other donors to make up the difference, by difficulties in obtaining adequate housing and meeting spaces for large and small gatherings that would be part of the events, and by an insufficient number of skilled people to solve these problems. The forum itself complicated matters, at one point voting to accept a construction firm's offer to renovate part of the municipal museum to be used as coordinating offices for the NGO events, and then voting to prohibit the firm from mentioning its having done so in any publicity. Not surprisingly, the firm walked away from the job.

Logistical problems notwithstanding, the conference took place, both in the official meeting at RioCentro and in the Global NGO Forum, set up along the beaches of central Rio de Janeiro city. After two years of preparations it passed in a blur, and was less significant for many of the organizers than the preparatory process. A multitude of displays, presentations, posters, events, and meetings at Flamengo Park and other locations around the city made the Global Forum a lively and energizing event for the eighteen thousand NGOs who came from all over the world to meet each other. These numbers included more than 4,500 Brazilian environmentalists and activists from other social movement organizations (*Who Is Who* 1992), who met their counterparts from other parts of the world, exchanged concerns and ideas insofar as languages would allow, and came out of the forum knowing much more about where they fit in a broader range of alternatives. They produced reams of documents, gathered in a CD-ROM of some eighteen thousand pages (*Earth Summit: The NGO Archives* 1995)—itself a real innovation in communications technologies for the time. In a precursor to the blogs of today, electronic publications like *Da Zi Bao* allowed ordinary participants to put down their thoughts and initiatives, and these have been preserved among many other documents. While the diversity of postings cannot be captured with just one example, a well-meaning Scot wrote, "Don't blame the Brazilians! All the chaos, delays and frustration for Northern NGOS are part of the reality and beauty of trying to find solutions in the South. We are just a microcosm of the world situation. Out of it comes creativity, understanding and hopefulness. Keep teaching us Rio!" During the course of the meetings, adherents of some urban social movements, union members, and other activists held demonstrations protesting against the governments' failure to contend seriously with the global inequalities that reinforced and indeed exacerbated domestic ones. For the world, this was an unprecedented outpouring of societal energy over two years, which created a model for other UN conferences and global negotiations in the 1990s (Friedman, Hochstetler, and Clark 2005).

For Brazil, whatever its strengths and limitations, the decision of the UN General Assembly to make it the host of the UNCED Conference precipitated a period of concerted identity formation and debate among Brazilian social movements and NGOS working on the environment, human rights, and development. Between mid-1990 and June 1992 the environment received more attention from the mass media and from government organs than in all the coverage from previous decades. Environmental organizations had to position themselves in relation to each other, to other social movement organizations with developmental and human rights agendas, to the Brazilian state, to foreign NGOS, and to both nongovernmental and governmental processes. For a set of organizations of which the vast majority were (and remain) small and local, this was a tall order.

It was a period of great possibility and of great stress. During intensive bursts of collaborative activity, it often seemed that more energy went into discovering the multiple differences among the organizations than went into the collaborative process itself. Where such differences—in ideas, in organizational capacity, resources, or scope of action—might have remained submerged in a discussion among environmentalists of, say, long-term strategies or relationships with party politics, they became much more apparent when the thematic range broadened, and especially when organizations had to take on practical responsibilities in a national—and international— organizing process. Procedural questions exacerbated differences, as the big professional organizations grew frustrated with the insistence of the small ones on a decision-making process giving equal weight to each organization, and their resistance to a more operational focus that allocated tasks according to different capacities.

Finally, Brazilian NGOs had to adhere to a schedule not of their own making: the preparatory stages for Eco '92 were punctuated with four international PrepComs, or preparatory commissions, for the official UN conference, alongside each of which NGO meetings took place to coordinate both NGO lobbying of official delegates and the preparations for a parallel conference (Friedman, Hochstetler, and Clark 2005). While Brazilian groups struggled mightily to come up with common substantive positions, foreign groups mainly wanted to know that housing and meeting space would be guaranteed. They were not particularly interested in how the Brazilians wanted to shape the conference discussion and did not even understand why they thought they should do so, given the gathering's international character. Foreign organizations wanted Brazilians to think more "internationally," while the Brazilians wanted to be recognized (and respected) for being Brazilian.

The Aftermath of 1992

The years after Eco '92 were a period of consolidation for environmental organizations in Brazil. The burst of energy that many had expected after Eco '92 was defused immediately after the conference by a broad-based social mobilization demanding President Collor's impeachment for massive corruption (Weyland 1993). The country hung on to every new revelation of the parliamentary inquiry commission, and people—especially young people—poured into the streets to demonstrate for his removal, painting their faces and making an enormous civic festival of the impeachment campaign. Environmentalists, many participating as avidly as others in this campaign, could not hope to compete with it for political space. Six months later Collor was out, and the energizing effect of the UN conference was partly dissipated. Still, for environmentalists and their organizations the run-up to 1992 and the conference itself were a watershed, after which the movement would not be the same. At both national and international levels, environmental activists had been able to define themselves in relation to other social and political interlocutors, sometimes changing both and sometimes reinforcing the prior beliefs of both. It was a remarkable period of political learning.

In the aftermath of 1992 the field of action for environmental organizations had changed. Although on the one hand the distance between the "King ONGS" and the smaller local organizations had narrowed and the mistrust lessened, the implications of choosing one path or the other were much clearer to all. Opportunities for collaboration with state agencies at all levels—federal, state, and municipal—had abounded during this period, as had the newer possibilities of collaborating with international organizations. The task of designing and implementing environmental education programs both in public schools and among the population at large increasingly fell to environmental groups. The arguments for professionalization became compelling for many organizations that had previously espoused a less formal, more voluntarist approach to militancy. Although the impulse toward professionalization came initially from organizations themselves, as with SOS Mata Atlântica (discussed above), after 1992 it was further stimulated by the administrative accounting and reporting requirements of the National Environmental Fund (originally envisioned by SISNAMA in 1981 and finally created in 1989 as part of the Nossa Natureza program), the few new corporate philanthropy programs that existed, and both European and North American donors.

Many environmental organizations wanted to use the forum's thematic groups to create standing networks that could—and did—support more sustained action on particular issues. The Atlantic Forest Network proved especially strong. In the years immediately after the UNCED conference, these thematic networks were the most visible remaining parts of the forum, although its members continued to meet occasionally when a more or less unified national voice was necessary. Thus, for example, they accompanied the UN's Rio+5 and Rio+10 (Johannesburg) conferences, producing joint statements on their visions of environmental sustainability (Fórum Brasileiro de ONGS 1997; Fórum Brasileiro de ONGS 2002). The forum has also been active in the World Social Forums held in Porto Alegre, forming part of the Brazilian organizing committee in 2005. And in recent years the forum has gained new life under the new acronym FBOMS. In 2004 funding from the Ford Foundation allowed FBOMS to acquire its first office of its own, after years of being housed in member organizations. It also has its own professional staff and has launched a new biweekly electronic newsletter in addition to a printed newspaper. The five hundred organizations which remain in FBOMS see a need for the organization to have its own political presence that can do ongoing political coordination (interview with Neuhaus and Bonilha 2005). In a sign of the original forum's penetration into national environmental politics and policy, the fifteenth anniversary celebration on 18 June 2005 brought together many founding individuals who were in the national ministries or congressional offices. In the eighteenth national meeting, at the end of 2004, four new working groups were added to the existing seven.

International actors—both foundations and environmental organizations—also became more visible within Brazil after 1992. International funding, previously limited mainly to traditional conservation activities and (after 1988) projects in the Amazon, became more widely available for organizations working on other ecosystems, urban pollution initiatives, and community participation in environmental activities. International development funding increasingly required the participation of "civil society" groups in discussion and implementation of projects. The Pilot Project for the Amazon sponsored by the G-7 countries, discussed in chapter 4, stimulated the creation of a spate of new environmental organizations in the region. Although some of these were formed essentially to get their hands on G-7 money, pseudo-organizations were more easily spotted and discredited than would have been true earlier. Finally, during the run-up to the Earth Summit, Greenpeace and the WWF established a significant organizational presence in Brazil, with support from their international secretariats, and Friends of the Earth (Italy) created a significant Amazon program. These organizations undertook campaigns that are intimately linked to international campaign strategies of their parent organizations. Brazilian environmentalists are no longer isolated.

Nonetheless, despite a number of important successes, and a gradual but steady growth in diffuse environmental consciousness and practices (environmental education in schools, recycling in urban areas, concern with water and air quality), environmental activism of the kind that had mobilized people through the 1980s and early 1990s did not appear to be growing. At a seminar held to assess the situation of the environmental movement in

1996, those present concluded that although the movement had not deteriorated, it had not really progressed—it was more or less as it had been before (Svirsky and Capobianco eds. 1997). Moreover, they concluded, the movement's leadership was not being renewed: the spokespersons of Brazilian environmentalism in the mid-1990s were the same people who had filled the role close to a decade before. (And the same people—now in their forties, fifties, and older—continued to do so well into the new century.) Part of this, the participants in the meeting reflected, was the result of professionalization. Although professional organizations mobilized many people to participate in projects or even campaigns, they did not recruit new ecological activists in the same way that they had done in the early days of the movement, when most organizations had dedicated themselves to the task. Thus they lacked a "base" in the sense that this had existed before. In fact, Capobianco argued, they were barely a social movement at all (Svirsky and Capobianco eds. 1997, 84–87). In one sense this was positive— environmentalism had become part of the program of many other kinds of organizations, as it should be. On the negative side, the number of activists for which environmentalism was the central commitment had not grown. Indeed, many former activists had burned out, or had retreated from intensive activity in the movement to support families and build professions.

To those reasons should be added one more, which is that the changing institutional context—fruit of the struggles waged by the first and second waves of environmentalists—presented a wider range of opportunities for effecting environmental change to new generations of Brazilians, or at least appeared to do so. Graduates of the new programs in sustainable development, ecology, and environmental studies at universities could look to a variety of roles in government, in business, or even as NGO employees that were not available to their predecessors. Real jobs (as opposed to the promise of future ones) were quite scarce, but environmentalism had become a sphere of professional activity, as much as—indeed more than—a sphere of political and social activism. Activists formed environmental organizations engaged in providing services—environmental education in schools, for example. However much some believed that this professional orientation diluted the idealism of the militant approach, others were sure that it would be made up for by the ability of the movement to spread much further into state and society—what Brazilians call its "capillarity."

Table 3.2 Associations for the Protection of the Environment and Animals, 2002

DATE OF FOUNDING	# OF ORGANIZATIONS	# OF PAID PERSONNEL
To 1970	17	49
1971–1980	66	870
1981–1990	226	751
1991–2000	968	1,263
2001–2002	314	73

Note: Excludes organizations disbanded before 2002.

Source: IBGE, Fundações Privadas e Associações, 2004, table 16; www2.ibge.gov.br/pub/Fundações Privadas e Associações.

Table 3.2 shows that Brazilian environmentalists were actually somewhat too negative about the developments in their movement in the 1990s. More than half the environmental associations in existence in 2002 were created in the 1990s, almost 20 percent after 2000. The newer organizations have fewer paid employees, on average, than their forebears, and those formed since 2000 have fewer still. Thus there seems to be a continuing base of volunteer activism among newer organizations (a suggestion supported by the figures cited in the first part of this chapter, most of which were compiled post-UNCED). On the other hand, the prevalence of comparatively large paid staffs among still-existing older organizations suggests that professionalization may be the key to organizational survival. This aggregate logic can be seen as well in the lives of individual activists, who find professionalization critical to their own participation as they move into new life stages.

A Life Cycle of Environmental Activism

The story of Rogério Rocco is illustrative. We met him earlier in this chapter as one of the young Turks of APEDEMA in Rio, outraged at what he saw as Pro-Rio's efforts to usurp the legitimate place of environmental organizations at the United Nations Conference. When we interviewed him in 1990 Rocco was in his early twenties, and devoting most of his time to building the environmental movement among youth. He had dropped out of high school two months before finishing, to spend more time on movement activities. He and others in the organization that he leads, Os Verdes, began as the student wing of the Green Party, the Center for Education (Núcleo de Educação), but quickly diverged from the leadership of what he called the big three—Minc, Gabeira, and Sirkis. Rocco thought that the founding group wanted to make the Greens into their personal club and refused to consider other views. The students organized a statewide student meeting; the Green leadership not only boycotted the meeting but expelled the group from the Núcleo de Educação. Around sixty people left the PV at that time. Rocco founded Os Verdes with one group of the dissidents, and at the time of our discussion the group had around fifteen members, mainly students.

After taking some time to find their feet, organizing demonstrations around a variety of national and local issues, Os Verdes continued to work mainly with secondary school and university students. In collaboration with other organizations of young people, like Grüde and MORE from Niteroi, Os Verdes set out to revitalize Rio's statewide APEDEMA. Young people themselves, they believed that youth were more open to alternative values and would recognize the need to change themselves in order to change society. Rocco worked at the time as an aide to a PT municipal councilor for Rio, Chico Alencar, who allowed him to use the office for environmental activities. In his role as a leader of APEDEMA, Rocco also became a vocal participant in the organizing process for the Earth Summit, vigorously defending APEDEMA's place in the forum's coordinating body in the face of IBASE's argument that the coordination should be done by organizations best-equipped to act professionally in organizing the events. During the early phase of professionalization, Rocco and Os Verdes argued that environmentalism would lose its vitality as a social movement if it developed a professional NGO identity.

When Chico Alencar was reelected for a second term as city councilor, Rocco resolved to go back to school. He redid his last year of high school and entered the law faculty at a university in Rio, graduating in 1997. He began working as an intern in the office of the environmental secretary of the Niteroi municipal administration (across the bay from the city of Rio de Janeiro), passing through most of the positions of responsibility in the office and ending up as secretary himself.

The 1990s were a period of redefinition for Os Verdes, as they were for many other environmental organizations. The optimistic expectations about social change that had accompanied political democratization in the 1980s collapsed during the continuing economic crisis and the growing hegemony of neoliberal discourses and policy. Politics became very pragmatic. Although there was much more space for environmental concerns in government, government officials wanted solutions, not protest. Organizations without professional structures lost credibility in the process. Many environmental organizations disbanded, while others decided to institutionalize. Rocco disagreed vehemently when Os Verdes's longtime ally Grüde decided to end an *ação civil pública* against a large project after a negotiation with the municipal government, PT politicians, the construction firms, and other parties. In addition to other adjustments, the agreement provided that a very large sum of money be set aside in the project for environmental education, to be administered by Grüde in exchange for its dropping the lawsuit. Although Rocco still considered Grüde an important and serious organization, and thought it had done well by becoming institutionalized, he wanted the actions of Os Verdes to retain a political edge.

By the end of the 1990s it was clear that unless the organization rethought its mission, it would have to disband. In addition to the generally demobilizing context of the decade, the life cycle changes that had occurred for the early NGO founders had begun to affect former youth organizers as well. Many of the organization's members had professional jobs in environmental fields, but no opportunity to use their professional skills in Os Verdes. It was time to professionalize the organization. As Rocco put it, "We were not adolescents anymore, living with our parents. In fact, many of us were becoming mothers and fathers ourselves. We had to make a living" (interview with Rocco 2004). Having begun Os Verdes with a statement of principles, the organization's members decided to develop a statement of means (Carta de Meios). They began by renting a headquarters, giving the organization an address and a phone number. They developed a two-pronged funding strategy: they would create a sustaining board made up of both individual contributors and firms, whose contributions could come in the form of services instead of money, and they would begin a mass membership campaign.

Before undertaking the campaign they established partnerships with such service firms as a printer of pamphlets and notebooks, a company making T-shirts, an advertising agency to do pro bono work for a campaign on the Atlantic Forest, and others. They were quite sure that they needed tangible inducements to offer to new members in addition to the usual training courses and the like. Further, they needed to have something to offer those who contributed larger amounts to sustain the organization—the ability to put their names on things that the organization produced, for example. As Rocco put it, "they will be making an investment and they will receive a product." This position contrasted starkly with the one taken twelve years before, when construction firms engaged in renovating space for NGO activities at the Earth Summit were prohibited from mentioning the job in their advertising. An example of the new kind of partnership was an agreement with a publisher to establish a book series, the first volume in which was an edited volume on environmental law, partially subsidized by the Heinrich Böll Foundation (of the German Green Party). Os Verdes received a certain number of copies to distribute, and the publisher had the right to sell the book through its normal channels. Os Verdes developed an introductory pamphlet and a web site.

Over time, the profile of Os Verdes changed. Although still pursuing political action—running for seats on environmental councils, helping to organize a national environmental conference sponsored by the Lula (PT) government, working in the Atlantic Forest Network—it pulled back from some other activities. Eventually APEDEMA hollowed out, comprising fewer than forty organizations. Rocco believes that it became much too partisan, reflecting both PT factional politics and the desire of current leaders to win influence by promising jobs in PT governments. Os Verdes also developed more professional capacity to undertake environmental projects on its own. As of August 2004 there were around thirty active members (including university professors in biology, administrators of ecological reserves, a sanitary engineer, and an energy planner) and another twenty who participated in the listserv but rarely appeared.

Although in 2004 Rocco recognized the difficulties in recruiting a new generation of activists, he believed that Os Verdes was doing the job better than most. Nonetheless, young peoples' expectations had changed. Today recent university graduates come to Os Verdes looking for a job, not for an opportunity to do volunteer work. For them environmentalism is a profession as well as a commitment, and they go away disappointed when told that the organization is not yet in a position to pay professional staff. During the 1990s and through the first years of the new century, jobs of any kind, even for trained professionals, were few. Still, it is encouraging that enough new people are joining the organization in search of a way to become involved. And like others whose careers we have followed, Rocco moved into government again, working at the ministry for environment, as the general coordinator of the National Environmental Fund, and as we write in mid-2006, as the executive director of IBAMA for the state of Rio de Janeiro.

Professionalization from Abroad

In the period after the Rio conference, professional NGOs also arrived in Brazil as branches of international organizations, with a history very different from that of their homegrown counterparts. Greenpeace Brazil, formally established in 1992, bore little resemblance to the small-scale local action groups, the larger social movements, or the domestically formed NGOs. It was set up after a long period of preparation, much as a franchise of a large firm would have been, with careful attention to the fit between local possibilities and the campaigns operating internationally. To understand how this worked, we need a bit of background on Greenpeace International.

Greenpeace has always had a distinctive internal culture, with an approach to environmental action based on civil disobedience, direct action, and high visibility. It was not just any environmental organization but a group of "warriors of the rainbow," a mythic combination of native American folk heroes and Borges's macho corsairs. "Greenpeace" is also a trademark, and to use it new groups must be duly licensed, and must fit the larger organization's image and approach to action. By the 1980s affiliates were governed by a powerful international office, funded by a percentage of the revenues of each national office and coordinating a variety of global campaigns.

Greenpeace has its members pay dues, and it functions as a cadre organization, in which paid professionals work with volunteers, some of whom eventually become cadres themselves. The organization has its own media specialists, fund raisers, experts in substantive areas, volunteer coordinators, and the like. Emboldened by its rapid success, Greenpeace adopted modern business management techniques at both national and international offices.

In the second half of the 1980s the money poured in; between 1986 and 1990 the annual budget of Greenpeace International doubled yearly. Anything appeared possible, and expansion to other areas of the world seemed like a way to increase the power of the organization's campaigns. In 1988 the organization began to plan expansion into Latin America.

Expansion into developing countries meant contact with different target populations and different political cultures. Unlike other Greenpeace offices, those in Latin America had virtually no prospect of becoming fully self-funding. Investment in creating the new offices was substantial (initially there were to be national offices in Argentina, Brazil, Mexico, and Costa Rica); the budget in 1989 was $728,000 (Greenpeace 1989), increasing to more than $1 million a year over the next several years. Besides regional and local organizers, the organization hired administrators, media specialists, and experts to prepare reports on possible campaign areas, particularly with regard to toxic wastes, ocean ecology, rainforests, and energy. The idea was to start off the new Greenpeace national offices with a solid base in research, strategically planned actions, and personnel.

Tani Adams,[18] a Guatemalan-American bi-national, became the coordinator of the Latin American expansion in July 1988 and was initially given a great deal of latitude in organizing the region, even to the point of agreement that the Latin Americans could choose the campaigns in which they would become active. This was unusual, as the international organization was at that time structured primarily around its campaigns, rather than around regions or nation-states. Adams believed that Greenpeace–Latin America should have a distinctive identity, drawn from the developmental path of the region and from the trajectory of the environmental movement in the region. Thus in Brazil she sought out activists who had long been active in environmental struggles to work with her in determining how Greenpeace Brazil would be organized.[19]

Greenpeace began to organize in Brazil when environmental activism and the visibility of environmental issues were both at the highest level they had ever been (or have been since). The campaign for the Constituent Assembly, the preparatory process for the Earth Summit, and the international uproar over deforestation in Amazônia all made it seem a propitious time for establishing an organization with Greenpeace's profile and resources. Romantic image aside, Greenpeace had a profile that made it quite different from the usual run of Brazilian environmental organizations. The oft-cited rules of campaigning laid down by the organization's international director, David McTaggert, were precise: "No campaign should be begun without clear goals; no campaign should be begun unless there is a possibility that it can be won; no campaign should be begun unless you intend to finish it off."[20] The process of deciding what campaigns to take on was protracted, since it required balancing local priorities with international interests. Greenpeace organizers had expected that the campaign on the toxic waste trade would be of interest to activists in Mexico and Brazil; however, they found that the volume of imported waste was tiny in relation to the problem of domestically produced toxins, requiring a shift in the orientation of the campaign (e-mail from Adams, 30 March 1991). Finances were extraordinarily complicated, in a period when Brazil and many other countries in the region were experiencing hyperinflation that devalued assets in local currencies almost by the hour. Further, no Brazilian environmentalist had experience thinking at the scale of activism that Greenpeace represented. Although its budget was not significantly higher than that of other large NGOs at the time, the ability of Greenpeace to focus on its campaigns, rather than having to subsist on project funding, gave it a great deal more flexibility than other organizations had.

Besides steep organizational and logistical learning curves, there was a political disjunction. McTaggert and much of Greenpeace refused to take political positions on non-environmental questions, even in the areas of human rights, women's rights, democracy, and social justice. In the Latin American context, this abjuration was all but impossible. In fact, the first director (briefly) of Greenpeace Brazil, José Carlos Libánio, was an indigenous rights campaigner. All of Brazil's Greenpeace organizers, including Adams herself, had come of age politically during struggles against dictatorships and human rights abuses. Throughout the organizational period, there was occasional tension between Adams and some European Greenpeace leaders over these matters, and over the degree to which the internal culture of Greenpeace ought to be affected by the incorporation of national organizations with distinct historical experiences. Adams took the position that as the organization became more truly international it must adapt; others believed that the culture was not negotiable—it was part of the organization's essence. Greenpeace, as one of the organization's leaders said during an e-mail debate, was not a social movement. It should not change with the context. Between 1988 and 1994 these debates shook the organization, becoming the proxy for other differences and playing out inside more focused debates over voting rights, the idea of a global energy scenario, the tuna-dolphin campaign, and the discussions that preceded the Earth Summit. These discussions about identity became all the more salient when after 1990 the growth in contributions slowed, not just for Greenpeace but for environmental organizations more generally. At the moment of their greatest ambitions, the leaders of Greenpeace had to begin scaling back.

Greenpeace Brazil carried out its first action in 1992, planting tiny white crosses in the area around the nuclear power plant at Angra dos Reis in Rio de Janeiro on the anniversary of Chernobyl, to memorialize those who had died in nuclear accidents. The organization was attractive to young people and undertook a variety of youth-oriented fund-raising activities. The Greenpeace Amazon campaign focused specifically on the export of high-value tropical hardwoods like mahogany, much of which went through illegal channels. Initially the effort was part of the more general Greenpeace tropical forest campaign; in 1999 the international organization mounted a campaign specifically focusing on Amazonia, and opened an office in Manaus (the Amazon campaign is discussed in chapter 4). By Brazilian standards the budget remained substantial. Over the past decade Greenpeace has become a highly visible environmental actor in Brazil, not only with regard to its actions, but also for producing serious, specialized studies and reports. It has appealed to young people, perhaps in part because of the glamour of its association with the developed North. Nonetheless, despite some occasional joint activities with other environmental NGOs, Greenpeace designs its activities according to a different logic and a larger scale, collaborating with Greenpeace cadres in other countries working on the same campaigns. Although to date Greenpeace Brazil has not met its goal of becoming self-supporting, by 2003 contributions from members amounted to around 33 percent of the organization's budget (Greenpeace 2003). In 1993 member contributions had been just 3.9 percent (Landim and Cotrim 1996, 13).

The 1980s were years of contradiction for Brazilian environmentalists. Great hopes were raised and dashed; opportunities opened up and dissipated again, or were found not to have been very robust in the first place. Toward the end of the decade environmentalists joined with other sectors of civil society to discuss how to host the nongovernmental events of the United Nations Conference on Environment and Development. As they responded to these and other political opportunities of the time—at all levels of politics, domestic and international—they engaged in open and extended debates about their own organizing strategies and political identities. Professionalization and a turn to socio-environmental discourses were two of the most important changes, creating a third wave of Brazil environmentalism that largely continues to this day.

The momentum leading up to the Earth Summit had a transformative effect upon the environmental movement generally, in three ways. First, it put environmental organizations into face-to-face contact—and continuous negotiation over principles and actions—with people from other social movements, in the context of the Brazilian NGO Forum. Second, it brought even quite small organizations into a web of interactions with environmental organizations the world over, giving them a sense of connection that had not existed before. Finally, it marked a clear differentiation between those organizations with the capacity to implement major projects on their own and those which worked more by marking principles, staking out positions, and raising consciousness. In other words, besides providing an international comparison set, the Earth Summit brought Brazilian environmental organizations into much closer contact with each other.

The process by which Brazilian environmental organizations mapped the global organizational field in which they played a part was an enormously useful educative process, a virtual apprenticeship in global environmental politics. As a result, more Brazilian environmentalists began to participate in international networks dedicated to global problems, rather than simply wait for foreign activists to work on Brazilian ones. Foreign environmentalists learned much more about Brazil; where before they had seen only the Amazon and its destruction, now they began to see environmental concerns and struggles related to a variety of ecosystems. Still, the Amazon remained Brazil's best-known and most compelling ecosystem, and both Brazilian and foreign environmentalists directed more and more programs to the region. But programs incubated in ecological logics had to adapt to the political realities of the region, which, contrary to the view from space, was amply populated and insisted on a voice. It is to these competing logics that we now turn.

4

Amazônia

When we speak of "the Amazon" or Amazônia, we are using a shorthand for an immensely diverse region whose geophysical, hydrological, ecological, and socio-historical characteristics have long interacted in complex ways. It became Amazônia through a political process of spatial organization. Drawing upon the work of Bertha Becker, we can say that this space was most visibly organized by exogenous forces—by the state, which built networks first of telegraph lines and then of roads to link the region to national and global markets and security apparatuses, replacing long-standing river networks. Less obviously, however, the region retained pockets of self-organization, with population groups that maintained a certain autonomy from the broader disciplining process. The area that the military regime designated "Legal Amazônia" (*Amazônia Legal*) now comprises nine states. From the outside, we see mainly what they share: forest, boom-bust cycles of resource exploitation, roads, and waves of migration. From the inside, highly differentiated social and political processes, ecological characteristics, and timing have produced different outcomes. Looking more closely at demographic patterns, we discover that almost 60 percent of the population of Legal Amazônia lives in cities, a fact that inverts most peoples' perceptions of the region (Browder and Godfrey 1997, 2). Beyond its ecological effects, urbanization has an undeniable bearing on the nature of local political organization.

Political struggle in the Amazon, for much of the last century, has been about realizing the economic value of the region's resources and taking advantage of and populating its vast spaces. The national policies that promoted these ends ignored the people who lived in the region when the Portuguese arrived, and over the last fifty years they have even ignored the existence of previous waves of migrants, who rather than settle in cities continued to depend on the extraction of forest products. As successive majorities rolled over the ones left behind in the last invasion, and a kind of social Darwinism held sway among those in power, protection of the forest became associated with protecting the region's minorities—people who pursue livelihoods which depend on the forest's continuing health: rubber tappers, indigenous peoples, brazil nut gatherers, artisan fishers, and the like.

By working with these groups, different sets of public authorities—the national environmental ministry and public interest groups—seek to claim space for a competing vision of regional development, with alternative networks of public and private, national, local, and international relations. The politics of protecting traditional populations has been an important point of entry into the region for external public authorities, but the cost of ignoring majorities is high in a democracy. For ordinary people working in agriculture, the lumber industry, commerce, mining, and the like, it can be difficult to understand why federal programs and foreign donors focus on small communities in the forest, thus opening themselves to the demagogic rhetoric of politicians opposing the donors' aid policies. Genuine protection of forest resources requires local support from public and private actors whose support carries weight with others. For all that the region's problems have been treated as socioeconomic, they are essentially political, requiring a struggle among competing visions of the region's future. More specifically, solving problems depends on state building, the creation of legitimate public authority strong enough to confront the lawless and predatory practices of those whose only interest is to grab as much wealth as possible in the shortest time. This is a tall order on such a territorial scale, particularly at a historical moment when state building is no longer in fashion. Under the circumstances, it is not surprising that extra-regional support should have been critical for struggles to protect forests in the Amazon, or that the definition of strategies should have been difficult. Even more striking is the role of contingent events, like the two murders that we recounted at the beginning of this book, in disrupting patterns just enough that determined activists can transform a momentary breach into a window of opportunity.

Understanding political interactions in the region requires understanding the institutional context of Amazonian politics—in which crucial actors are not merely environmentalists and policy makers in the abstract but municipal, state, and national governments—and the interactions among the actors as well as the other influences upon them. During the late 1980s and 1990s the developmental model was in manifest crisis, and the vacuum created by the politics of state absence was filled by privatized violence committed by local bosses or growing criminal networks. Nonetheless, this was also a period when those resisting the dominant model developed alliances, information networks, and linkages to other parts of Brazil and the world. By accelerating the flow of information and resources through these networks, activists hoped gradually to enlarge the space available for endogenously generated development alternatives. More recently the Cardoso and Lula governments made efforts to resume a centrally defined policy process for the Amazon, even as parts of their own governments did what they could to either prevent this process or mitigate its effects. Although they are still much weaker than the forces they challenge, the resisters have taken their struggle inside the state itself, and challenge it from within as well as without.

Geography and Geopolitics of the Amazon

Amazônia is immense—in size, mineral wealth, biodiversity, and freshwater, and in its hold on the imagination. It occupies one twentieth of the earth's surface, two-fifths of South America, and three-fifths of Brazil, and has one-fifth of the earth's supply of freshwater. It includes parts of the territories of seven countries, but Brazil's 63.4 percent is by far the largest share (Becker 1990, 9). The colossal and diverse visions of the Amazon that still hold sway in national politics profoundly affect the region's politics: for all the exploration and settlement, for all the exactness of satellite imaging, the Amazon remains a vast repository for the grandiose dreams of individuals, private firms and other economic actors, and the nation-state itself. It has aroused the imaginations of enough illustrious foreigners to make Brazilian suspicions of foreign designs on the region understandable, if not always accurate. Nationally, it has served as a potent symbol of Brazil's potential power. If Brazil was *o pais do futuro*, the Amazon basin was its *paisagem*, its ticket to being the country of the future. It was an imaginary landscape of wealth there for the taking—a "land without people for people without land," as the Brazilian military said, whose indigenous peoples and other local populations were quite simply painted out of the picture.

Just as Amazônia's economies have known boom and bust cycles, periods of geopolitical significance have alternated with periods of relative neglect. In colonial times the region was explored by travelers seeking medicines and other products for the European market. During the nineteenth century soldiers and missionaries moved into the region, relocating indigenous populations and proposing to "civilize" them through evangelization (Procópio 1992, 155–235). From the late nineteenth century to the early twentieth there was a boom in the extraction of latex from rubber trees to feed the industrializing economies of the United States and Europe (Weinstein 1983; Dean 1987). The boom ended when the British, who had taken rubber seedlings from the Brazilian Amazon to Kew Gardens, successfully produced latex on plantations in their Southeast Asian colonies. This British perfidy made an early contribution to a continuing narrative about foreign designs on the region's wealth (*a cobiça internacional*) that fixed Amazônia's place in the discursive construction of Brazilian nationalism (Reis 1982). At the beginning of the twentieth century the region was linked to the rest of Brazil by the telegraph, and many of its territories were demarcated by the mission led by Marechal Cândido Mariano da Silva Rondon, just as the rubber boom was ending. In the 1920s and 1930s Amazônia became an agricultural frontier and area for mineral prospecting.

The United States funded a brief resurgence in the rubber economy, to supply Allied needs for rubber when the Japanese occupied Britain's Southeast Asian colonies, but it ended its funding in 1947. In 1953 the developmentalist government of Getúlio Vargas set up the Agency for the Valorization of the Amazon (SPVEA), with a special fund designated in the Constitution of 1946 to finance development projects there. Under Kubitschek the Belém-Brasília highway, connecting the Amazon's main port city with the newly built national capital, became the first major ground link between the mouth of the Amazon River and southern Brazil. Gradually, over the next decade, migrants began to try farming in the region, and many went to seek their fortunes by prospecting for gold and other minerals (Foweraker 1981; Schmink and Wood 1992, 49–53).

The generals who took power in 1964 saw the region in geopolitical terms, laid out lucidly in General Golbery do Couto e Silva's *Geopolítica do Brasil* (1967). In a global battle between barbarism and civilization, national integration was a paramount condition for the defense of the civilized West.

The enemy was as likely to come from within as from without, and the role of the armed forces was thus to establish not only secure borders but also a secure and developed society guided by a national security state. The Amazon basin—sparsely populated and of difficult access but a potential repository of unimaginable riches—could only become impregnable to subversion if settled and developed economically. The geopolitical vision drove public policy from the 1960s to the late 1980s, with little consideration for the Amazonian ecosystems and their varied occupants. The generals were determined to develop, populate, and extend their institutional control over the largely unmapped and unincorporated Amazon.

In 1966 they established an Agency for Amazonian Development (SUDAM) to implement a web of fiscal incentives, subsidized credit, and colonization packages which brought both small settlers and large businesses to the Amazon region (Bunker 1985; Mahar 1989; Binswanger 1991; Almeida 1992; Redwood 1993). On SUDAM's eighth anniversary the agency published a document which approvingly cited former president Getulio Vargas's vision of the Amazon: "To see Amazonia is the heart's desire of the youth of the country.... To conquer the land, and tame the waters, subjugate the jungle, these have been our tasks. And in this centuries-old battle we have won victory upon victory. The Amazon, under the fertile impulse of your will and your labour, will not be merely another chapter in the history of the earth, but, like other great rivers, will become a chapter in the history of civilization" (SUDAM n.d.).

To open the region for mineral extraction and industrial development (exemplified in the Grande Carajás program), the military government began an extensive program of road construction (Becker 1982; Becker 2001).[1] The idea was to integrate Amazônia into the national territory, make its vast resources accessible for economic development, and secure political control over territory, sealing it off from developments in neighboring countries, to prevent the establishment of revolutionary *focos* in the region. Roads would bring in more settlers and make possible large investments in infrastructure like the Tucuruí Dam. The roads brought changes to the region, valorizing manufacturing, especially in Manaus but also in Belém and Santarém, and mining in Rondônia, Amapá, and especially Carajás. Carajás is an area in the south of Pará of around ten thousand square kilometers, surrounded by mountains four to six hundred meters tall. In 1955 Bethlehem Steel had begun to export large quantities of manganese from the region, joined in 1966 by Codim. Then in 1967 U.S. Steel opened up an extraordinarily rich vein of iron ore. At the end of the 1960s the military government pushed U.S. Steel into an uneasy association with the Brazilian parastatal Companhia Vale do Rio Doce (CVRD), which had been formed in the 1940s to mine iron ore in Minas Gerais. The association lasted until 1977, when U.S. Steel pulled out, receiving a large indemnity for doing so (Hall 1991; Roberts and Thanos 2003, 150–51).

The Carajás story, and especially the relationship between CVRD and U.S. Steel, illustrate the ambiguities in the continuing contest over the nationalist mantle as played out in the Amazon. Since the military government saw iron ore as a critical resource from a national security standpoint, it did not want to leave it entirely in the hands of a U.S. company, and it forced U.S. Steel to accept state participation in the form of the CVRD. Nonetheless, because its national security doctrine was deeply anticommunist and thus—in geostrategic terms—pro-American, the military government welcomed multinational investment, and contributed to the creation of an enclave in Carajás (Becker 1982).

The project's opponents also appealed to nationalism, condemning the alienation of strategic resources, and the export of unprocessed ore at low prices for the use of foreign companies which would compete with Brazilian industries (Pinto 1982; Valverde 1989). The conservation organization CNDDA (discussed in chapter 2), whose founders included military officers purged from the armed forces for their political views after the coup of 1964, was a leading proponent of this left-nationalist position.[2] The CNDDA was founded in 1965, in reaction to a set of reports associated with the Hudson Institute proposing an extensive system of "great lakes" and hydroelectric dams in the Amazon to satisfy the energy needs of the hemisphere. These proposals were interpreted at the time as expressing U.S. government policy, and many Brazilians still believe them to have been so, although we have never been able to find real evidence to that effect (Keck 2002, 47).

Over the next decade the composition of the agricultural frontier also changed. Between 1965 and 1974 subsistence farmers along the Belém-Brasília highway were expelled to make way for enormous cattle ranches, whose pastures required the burning of huge swaths of forest. Many of the early small-farmer settlements organized by the federal Colonization and

Land Reform Institute (INCRA) in Pará state failed quickly, causing their settlers to join others in pushing back the frontier, only to be expelled by force from their new lands by the hired guns of land grabbers from Minas Gerais, São Paulo, Paraná, and Goiás (Bunker 1985).

Starting in 1974 the Geisel government embraced an agribusiness model of large ranches and other agricultural enterprises that soon made Amazônia a major beef exporter. Not only individuals but also industrial and service firms from the South took up ranching during this period—Volkswagen, Varig, and others. Workers, lured to the region with extravagant promises of future wealth, cleared forest to make pasture and were let go to make their own way. They built new towns and found temporary work, and many got caught up in the gold rush of the 1980s. Land struggles in the 1970s and early 1980s grew increasingly violent, especially in the south of Pará, the west of Maranhão, and the north of Tocantins (Arnt and Schwartzman 1992, 104).

The transition to democracy did not improve matters; if anything they worsened, as military control gave way to a power vacuum that unscrupulous actors were eager to exploit. Bertha Becker counts 1985—when civilian government was restored in Brazil—as a pivotal year for the Amazon in two respects: it marked the exhaustion of the national-developmentalist model propelled by state intervention in the region's economy and territory, and the formation of the National Rubber Tappers' Council, along with the beginning of an endogenous push to develop an alternative model based on local needs (Becker 2001, 141). We discuss the endogenous process later in this chapter; here we are considering the collapse of the exogenous one.

During the 1980s the Amazon, especially Rondônia, served as an escape valve for displaced farmers and opportunity seekers from other regions. The process by which the state provided incentives for appropriating and valorizing space was not strong enough to ensure its formal ordering. As migrants, entrepreneurs, opportunists, and scoundrels flooded the region, the absence of institutional authority and especially of a regulated system of land tenure rewarded the strong and subjected or silenced the weak. Cattle ranching expanded, taking advantage of a stipulation in land law that equated land clearing with effective occupation, an aberration not remedied until the 1990s (Mahar 1989). The cycle of predatory practices, exemplified in the annual "burning season" in which vast swaths of forest were cleared for pasture or farmland by burning, proved hard to control. Although this was partly due to a lack of appropriate personnel, information, and equipment, the bigger problem was political. State and local governments in the region had no desire to stop the practices that were causing such consternation elsewhere, and Brazil's constitutional structure gave them ample political resources with which to resist.

Institutional Context

When foreigners decry the continuing destruction of the Amazon forest, they generally end with an exhortation that Brazil must *do* something about it. The exhortation stems from three false assumptions, along with an implicit belief that the system of state planning and investment in regional infrastructure and development incentives that characterized the military period remains in force and can be redirected toward constructive ends, when in fact the model has already collapsed. The first false assumption is that there is a unitary public authority, in contravention of the constitutional division of power and revenues among federal, state, and municipal governments. Second is the assumption that there is a single policy, or at the very least a keystone policy, regulating or at least capable of regulating the seemingly relentless process of deforestation. And finally, there is an assumption that the state (at some level) is effectively present and capable of determining the public interest and exercising authority in furtherance of it.

Reacting to an authoritarian government by means of which the military had drastically centralized power and resources, the writers of Brazil's Constitution of 1988 strengthened its federalist components, mandating redistribution of a large portion of tax revenues from the federal government to states and municipalities. Ever since, budget deficits have alternated with periods of fiscal austerity, producing complicated bargaining between federal and state governments (Samuels 2003). Besides exercising power in their own jurisdictions, state governors generally have a lot of influence with, if not control of, their states' congressional delegations. The overrepresentation of the northern and northeastern states in the Brazilian National Congress gives them added influence when negotiating votes for or against federal government programs (Selcher 1998).

Presidentialism in a multiparty democracy confronts Brazilian presidents with a perennial problem of constructing a majority coalition to pass their programs. This is a costly process, its price generally paid in control over federal jobs and constituency goods, and often paid over and over again. The model of a "politicized state" proposed by Douglas Chalmers in 1977 remains strikingly relevant today in describing Brazilian national politics. Because of the overrepresentation of smaller states, their deputies are courted especially diligently, and often the state governor is an important broker for their votes. The Brazilian Constitution sets the minimum number of deputies for a state at eight and the maximum at seventy. Vast differences in population between the smallest state (Roraima) and the largest (São Paulo) skew the vote quotient considerably, such that Roraima has one deputy for every 40,549 people and São Paulo has one for every 529,034. In 2002 the seven states of Legal Amazônia had twenty-five more seats in the Chamber of Deputies than they would have had were representation strictly proportional; the southeast had thirty-nine fewer, and São Paulo in particular had forty-two fewer (Soares and Lourenço 2004, 118). This arrangement gives states in the Amazon a great deal more political weight per person than other states in Brazil have.

In bureaucratic terms, federal organs often operate in the region through *convênios*, or subcontracts, with state government organs; for example, there are subcontracts between IBAMA and state environmental agencies, and between INCRA and state colonization or land agencies. States can draw up zoning plans, but municipal authorities must accept and enforce them. Most enforcement falls under the auspices of military police (controlled by state governments) or the municipal civil police, both of which are badly trained, badly paid, notoriously corrupt, and often engaged in illegal activities themselves. Indeed, with few exceptions the Brazilian judiciary plays a key part in supporting the powerful over the powerless. The Ministério Público discussed in chapter 1 as a new enforcer of environmental laws is much weaker in the Amazon than in the South. The state Ministério Público has been nearly absent in Pará, for example, leaving most fights to be waged at the federal level (McAllister 2004).

Besides federalism, policies and actions affecting deforestation are also spread among a wide array of ministries and agencies, as well as private actors. Planning and regional development ministries and state secretariats give out incentives for investments resulting in the kinds of changes that environmental ministries and secretariats try to regulate and limit. Policies designed to improve the balance of payments by expanding export agriculture stimulate further deforestation. The maintenance of a high primary surplus to meet IMF requirements results in the nonrelease of budgetary funds needed to monitor conservation areas. Interministerial environmental planning groups include the people from each ministry most interested in the environment, rather than those with the power to make decisions.

Very few public officials in any sector, or at any level of government, want to be posted outside the state or national capital: the Amazon seems more manageable from a distance. They are not alone in this—NGOs fall prey to the temptation as well. When the WWF came up with a proposal to preserve 10 percent of the Amazon, its representatives visited the Ministry of Environment and Legal Amazon with a map of the areas they believed to have the greatest ecological value, and were adamant that the parks to be included in the 10 percent should have no people in them, as if there were no one there (Allegretti 2005; interview with Lourenço 2002).

The problem is rarely the laws, or at least not their intent. Use of natural forests is regulated with environmental impact assessments; burning and clearing permits; cutting restrictions specific to the property; cutting restrictions specific to the locale; forest management requirements; and restrictions on export of forest products (Lele, Viana, Verissimo, Vosti, Perkins, and Husain 2000, 20–21). The Forest Code of 1965 was highly protective of forests, requiring in the Amazon region that between 50 and 80 percent of any property be maintained uncut, in addition to which there had to be vegetative protection of riverbanks, slopes, and other fragile areas. In 1988, with the implementation of the Nossa Natureza (Our Nature) program under Sarney, fiscal incentives for ranching investment in the Amazon were suspended and monitoring of forest burning was stepped up, though it remained much weaker than the situation warranted. Withdrawal of fiscal incentives was no match for the growing profitability in the 1990s of the ranching sector, and medium- and large-scale cattle ranching remains the leading cause of deforestation in the region (Margulies 2003), although soybean cultivation is advancing as a major cause of land-use change (Hecht 2005). Provisional measures and decrees in 1996 toughened the requirement for maintaining forest cover from 50–80 percent of total land to 80 percent of the forested part of the property; where approved zoning plans existed, smallholders were not subject to this restriction.

The issue of timber is more complicated. Over the last twenty years, as the exhaustion of the Atlantic Forest put increasing pressure on the market for timber products, logging in the Amazon became more profitable and expanded. The regulations, as noted above, are restrictive more than regulative; there is no strong incentive either for firms themselves or for local enforcement officials to comply. Short time horizons cause local politicians to ally with loggers, and in some areas loggers have been elected to local posts (Lele, Viana, Verissimo, Vosti, Perkins, and Husain 2000, 23). Many local politicians protested when a decree in 1996 put a two-year ban on new management plans for high-value timber while IBAMA investigated existing concessions (22). Enforcement agencies are grotesquely understaffed, and in a number of cases enforcement officials have been found to be cooperating with illegal logging; in a joint operation against illegal logging in July 2005 conducted by the Ministério Público, the Federal Police, and IBAMA, IBAMA's chief official in Cuiabá was arrested along with forty IBAMA employees in Mato Grosso and Rondônia (*Folha Online*, 19 August 2005). With short-term licensing and lax enforcement (or active connivance), loggers have no incentive to change their practices.

In 1977 SUDAM proposed to establish a network of national forests managed for timber concessions; although interest among investors was apparently quite low (Garcia 1987), the measure was one of several against which environmental movement organizations organized the "Risk Contracts" campaign (discussed below). The term was a misnomer, taken over from a controversial measure allowing foreign companies to prospect for oil in Brazilian territory; timber concessions are not risk contracts, as the timber is obviously either there or not. SUDAM proposed to grant long-term forest management contracts to private (including foreign) investors, who would be required to follow guidelines set by the government. The length of contracts, one of the elements to which opponents would object, was a recognition that without a long-term contract the incentives for reforestation and for careful forest management disappeared; the firm would only be interested in rapid extraction of the most valuable timber.[3]

At the time, environmentalists were more worried about trying to block the increasing private and especially foreign exploitation of forest areas than they were about creating a legal framework to regulate eventual logging activity. Logging became a pressing matter only in the late 1990s. The new forest code now under consideration in the Brazilian Congress represents a substantial improvement, granting concessions long enough (forty years) to provide an incentive for stewardship, and making it much more feasible to distinguish between legal and illegal logging operations. The law would also create a forest service, removing the actual monitoring function from IBAMA, which has not been able to handle the job. Nonetheless, the agency would retain responsibility for having granted the concession, and bears the blame, along with the private logging firm, for abuses should they occur (ISA 2005).

The Politics of State Absence

The problem of law enforcement in the Amazon goes well beyond the issues of environmental crimes and the weak institutional capacity of the environmental agencies charged with enforcement. At its core, the problem is one of political will—in some cases the lack of desire to expend the necessary political capital and resources to enforce the law, and in others the active desire to prevent its enforcement. In the words of a prominent journalist from the region, "You get the idea here [in Pará] that *poder público* [government] is just a front for groups set up to loot the public purse and use state power as a tool for exploiting Amazonia's natural resources. These days, most of the Amazonian elite is a byproduct of organized crime. There is a vast power structure. The groups which grew out of land grabbing and clandestine exploitation of natural resources have much more power than the state does. That's why there is this shocking kind of violence. Criminality has turned Amazonia into an enormous green Sicily." Lúcio Flávio Pinto (interviewed in *Jornal do Brasil*, 21 February 2005).

Pinto himself can attest to the criminality of the region. A journalist from Belém who for the last seventeen years has published his *Journal Pessoal* on the region (modeled on *I. F. Stone's Weekly*), he has been beaten up, received death threats, and been sued for libel close to a dozen times for exposing the actions of corrupt officials and businesses in Amazônia. A recent libel suit demonstrates the impediments that he must constantly overcome. The suit claimed that the following sentence from an article he had written drew undue conclusions instead of stating facts. Referring to land grabbing in the region, Pinto wrote about the plaintiff, "Cecílio do Rego Almeida is only the most audacious, clever, and well connected of these land pirates." He was not the first to note it. In a white paper on land grabbing in 2002, the federal Ministry of Agricultural Development referred to Almeida as the perpetrator of the most serious attempt to usurp public lands in the country. However, in an irregular proceeding during which the regular judge was replaced by another exclusively for the one case, Pinto was found guilty and made to pay 8,000 Brazilian reais in punitive damages, with interest and monetary correction dating from when the suit was filed in 2000.

The period of the democratic transition coincided with a speeding up of the process of occupation of Amazônia that the Médici government had initiated a decade earlier, with the influx of ranchers, land speculators, small farmers, miners, construction workers, and lumber companies. Some came to make a quick buck and leave; others came to stay, the last stop on a long chain of migrations. Some say that when the military regime was in power, the pervasive militarization of the Amazon appears to have kept vigilantism under control, and to have prevented other groups from developing their own "violence machines" able to act with impunity in the region.

At the same time, the "war on drugs" conducted with Washington's sponsorship in Colombia and Peru was moving cocaine trafficking elsewhere, especially to the Brazilian Amazon. Cocaine entered Brazil through a variety of means: trade across the river with Bolivian drug groups by gold miners near Guajará Mirim in Rondônia, and clandestine airstrips built all over the region that linked large landowners to drug sources in Colombia and arms dealers in Suriname. The spread of small-scale trafficking through migrant networks contributed to the rapid development of the national market for cocaine in Brazil. To buy cocaine, systems were established to generate desired trade goods—car theft rings, for example, as well as a variety of smuggling and money-laundering operations (Geffray 2002b).

The situation varied from one part of the region to another, but in all cases migration to the region was met by notable failure of the state to perform crucial services like registering landholdings, resolving conflicts, providing extension services, and building adequate infrastructure. Rural violence was endemic, as big landowners expelled smaller ones from their holdings under threat of violent death (carried out sufficiently often as to be credible), hundreds of rural lawyers and union organizers were assassinated, and a class of professional enforcers became a familiar part of the landscape. In some regions, especially southern Pará, levels of violence rose very quickly from other causes as well—mines closing down, leaving mineworkers stranded; gold miners struggling to defend their claims or scrabbling in open-pit excavations like Serra Pelada; and criminals engaging in myriad clandestine activities, particularly drug trafficking and the illegal extraction and smuggling of mahogany and other timber.

There was a great deal of money to be made in the region (especially if you were not overly concerned with the legality of how you made it), both for entrepreneurs and for politicians who provided cover for them. Under the circumstances, the likelihood of prosecution for any of these activities was extremely low. The impunity with which people could operate outside the law made possible an expanding power structure of which criminality was an integral part. In some cases members of criminal networks coopted politicians and organs of the state or ran for office themselves, and in some cases politicians and state employees, seeing opportunities, took over criminal networks. In his studies of drug trafficking and the state in the region, Christian Geffray identified the key move: officials (or the state as a whole) would "renounce the exercise of their duties in the struggle against drug traffickers while retaining their position—since their resignation would serve no purpose. These officials' act of *renunciation*, coupled with their *failure to relinquish their office*, is at the very core of the corruptive transaction" (Geffray 2002a; emphasis in original).

Thus "institutional weakness" and "absence of the rule of law" often cited by studies of the "failure" to enforce environmental standards or pursue miscreants is not an accident of recent settlement but rather a strategy deliberately pursued by powerful operators in the region for which a more robust state geared to maintaining law and order would be highly inconvenient. As Anna Tsing brilliantly demonstrates for Kalimantan, Indonesia, the frontier is a project: "The frontier is made in the shifting terrain between legality and illegality, public and private ownership, brutal rape and passionate charisma, ethnic collaboration and hostility, violence and law, restoration and extermination" (Tsing 2005, 33). Frontier making can get out of control, however, and become crisis. Violence ceases to be a distraction from the story of profit and accumulation, and becomes the center of the landscape: "Everyday processes of frontier-making become crises when, in contingent concert with regime disintegration and international 'loss of confidence,' frontier violence and destruction take on a new magnitude . . . Chaos: a frontier spun out of control, its proliferations no longer productive for the authorities" (Tsing 2005, 42–43).

Even when state and local governments in the region have not been overtly criminal, they have been almost by definition boosters of colonization and development within their jurisdictions, with few (and recent) exceptions. There is a considerable difference between the Amazonian states of recent colonization and those with older population centers like Amazonas and Pará, whose capitals, Manaus and Belém, house distinguished scientific research institutes, strong universities, and important intellectual communities. Nonetheless, just as the sources of oppositional social capital are stronger in these locales, the governing elites are correspondingly stronger as well. Thus the main arenas for raising demands have been the press, universities, at times the Catholic church, the federal government, and both intergovernmental and nongovernmental forums in the international arena.

Resistance

There have been three periods of resistance in this prevailing story of Amazonian politics. Although Brazil now boasts many strong supporters of environmental stewardship, when they began, Mary Allegretti claims, there was almost no one on their side (Allegretti 2005). Confronting the problems of the region has always required allies; environmentalism in the Amazon has always involved linking actors who viewed the region from different scales, from different perspectives, and from both inside and outside its territory.

Three sets of domestic actors tried early on to stem the tide of destruction in Amazônia; they set the stage for the moment when the region could claim the world's attention. Brazilian scientists worked with international counterparts in the 1970s to map the astonishing complexity of the Amazon's ecosystems, turning to advocacy politics toward the end of the decade as they grew increasingly concerned about the impact of the military government's policies for the region. Some joined a second moment of resistance in the 1970s. A national environmental and antimilitary campaign against the government's plans for forest development in 1978–79 temporarily mobilized many Brazilians across this gigantic country. During roughly the same period, the rubber tappers' leader Chico Mendes and other regional visionaries were creating strong organizations in the western Amazon, in the state of Acre; while their struggles did not define themselves as environmentalist, they proved to be the crucial domestic link that made subsequent transnational campaigns powerful—where they existed. This marked the first time that people native to the forest itself became the chief advocates in the struggle for its conservation.

International actors played an important role in this process. Transnational linkages were built between international environmentalists on the one hand and rubber tappers, indigenous groups, and their allies on the other, who brought external resources to bear on an internal problem: the need to ensure their rights to land. They persuaded powerful actors, including members of the U.S. Congress and the Treasury Department to put pressure on the World Bank, which in turn would put pressure on the Brazilian government. Thus these activists precipitated what Keck and Sikkink called a "boomerang" pattern of influence: they blocked or stalled funding for several development projects in the Amazon until demands for environmental protection and indigenous rights were addressed (Keck and Sikkink 1998a).

That version is correct, but is akin to a view of the forest from above the canopy. Here we delve deeper into the complex moves in this struggle, as the Brazilians and their advocates sought favorable venues among multiple levels of governance, and developed new channels of information and influence. Well before intergovernmental channels and multilateral lending agencies became involved, allies included the Catholic Church, the rural labor movement, the Workers' Party, international environmental, development and human rights organizations, and a variety of others.

In the aftermath of the transnational moment, Mendes's murder had the effect of a centrifugal force, catapulting onto a national stage, and into political spaces that had been suddenly opened up, many of the people who had been associated with him in the struggles of rubber tappers in Acre. We consider the trajectories of some of the domestic actors set in motion in Acre: Tony Gross (profiled in chapter 3); Mary Allegretti, anthropologist from Paraná; Marina Silva, rubber tapper, university professor, senator, and minister of the environment; and Jorge Vianna, who helped to develop an extractive reserves proposal and became governor of Acre. International influence and actors continued to be present in the 1990s and after, but they took on a new set of roles. Most notably, as Becker (2001) points out, the changes of the 1990s enabled the construction of a new kind of network in the Amazon, one providing a counterweight to the network of roads and physical infrastructure that unleashed so much destruction: a network of information exchange, electronic communication, and alliances stretching to the far reaches of Brazil and the world.

Early Brazilian Resistance to Destruction of the Amazon: Science, Scientists, and Research

Amazônia is a complex place comprising multiple ecosystems (Borelli et al. 2005). There is no cookie-cutter conservation plan that would ever be appropriate for the entire region. Although research is constantly yielding new information, a great deal is still unknown about the larger impact of small changes—deforestation of a small area might have a large impact on the reproduction of particular species, whereas deforestation of a larger area might have a smaller impact. There are areas of soil that would support agriculture and many areas that would not. Human beings have always caused changes in the forest, but only in the last half-century have they had the technology and resources to affect it so profoundly. The arrival of large-scale entrepreneurship—in the form of multinational mining consortia, and big ranching and logging companies—changed the landscape, producing a patchwork of devastation. The long-term impact of this devastation is still unknown, but it might extend to climate and precipitation patterns far beyond Brazil's territorial boundaries. These are just a few of the puzzles which have made the Amazon such a compelling and fruitful location for scientific research.[4]

Amazônia has first-class research centers of long standing. The Museu Paraense Emílio Goeldi in Belém was established in 1866 to do scientific research on the region. The Institute for Research on the Amazon (INPA) in Manaus, with field stations throughout the region, was set up by the Vargas government in 1952 as part of a nationalist reaction to a UNESCO project, proposed by a Brazilian chemical engineer, to establish an international scientific research institute in the Amazon, the International Institute of Hiléia Amazônica (IIHA).[5] In fact the INPA proposal mirrored aspects of the IIHA proposal, and several of the same people were involved (Maio 2005, 123–26). Brazilian scientists had plenty of international allies from the outset, in that the scientists who lived and worked in the region included many expatriates, and other foreign researchers spent as much time as possible there. At the same time Brazilian scientists and conservationists were integrated into an international community of their peers.

In the 1970s scientists began to engage in advocacy, concerned (justifiably, as it turned out) that the Médici government's drive to settle the region would produce a rash of ill-conceived and ill-supported colonization schemes, destroying important swaths of forest and species habitats in the process (Keck and Sikkink 1998a, chapter 4). Scientists made common cause with anthropologists, who were concerned above all with the devastation that mineral exploration was wreaking on indigenous lands, which were opened up for mineral exploration by Decree no. 88,985/83.

During this time the Amazon was still very far away for most of the environmentalists in the south of Brazil, who were mainly concerned with the pollution they experienced every day. Similarly, "movement environmentalism" of the kind that developed in the South during the 1970s and 1980s was present mainly in large cities like Belém and Manaus, and although urban environmentalists spoke of protecting the Amazon, they did not know a great deal about it. Bridges were provided by scientists like the geographer Aziz Ab'Saber, a renowned paleoecologist and paleoclimatologist of the Amazon at the University of São Paulo, who was also active in the São Paulo environmental movement and in the Commission to defend the Community Heritage (CDPC), discussed in chapter 2.

The "Risk Contracts" Campaign

After the political liberalization initiated in 1974 produced a freer press and elections in that year increased the number of opposition MDB deputies in Congress, members of the opposition tried hard to penetrate the secrecy with which the military still surrounded policy making related to the Amazon. Throughout the period of political opening, Amazônia remained high on the military's national security agenda. One of the protests against the secrecy of regional development policies was a national campaign in 1978–79 against a proposed new timber concession policy. This campaign was generated from outside the region, involving newly mobilized environmentalists and other opponents of the military regime. They called this a campaign against "risk contracts," associating them in the public eye with the oil exploration "risk contracts" granted by President Geisel in 1975 which had aroused public indignation. SUDAM called the project Regional Yield Forests (Florestas Regionais de Rendimento). At the end of 1978 reports of new government plans for timber harvests in the Amazon leaked to the press, newly freed from military censorship. The plan provided for the sale of concessions for huge tracts of the forest to investors, who would then systematically develop the Amazon's forest products potential (Garcia 1987). Even without a full restoration of political and civil rights, the plan created an agitational niche (Keck and Sikkink 1998b, 223).

Proponents focused on the claim that proceeds of the sales would equal half the national debt, clearly a desirable goal. They were therefore caught off guard by the size and vehemence of the opposition. A youth sector of the opposition MDB party in Amazonas state organized the first protests at the end of 1978. Its Movement in Defense of the Amazon eventually had local committees in eighteen states and the federal district. Around fifteen hundred people reportedly attended public debates on the Amazon in São Paulo. The CDPC, based in São Paulo and by that time an eighty-member coalition of environmentalist and related groups, joined in. Some CDPC members tried to form a group to work exclusively on the Amazon, but this effort proved unsustainable (interview with Ab'Saber 29 April 1991). Lutzenberger was one of the movement's leaders, and he was frequently cited in the extensive newspaper coverage of the mobilizations. Using protest tactics and institutional channels, participants held numerous demonstrations, letter-writing campaigns, and conferences to educate themselves and local populations, and they lobbied the congressional inquiry committee that was eventually created. The CDPC also planned (but never brought) a lawsuit and gathered signatures in a petition drive.

The participants in the Movements in Defense of the Amazon exchanged documents nationally and coordinated their mobilizations. They organized the First (and only) National Congress in Defense of the Amazon in Brasília in October 1980. Statements from the São Paulo and Acre branches made three arguments: the government's plan encouraged the growth of enormous foreign and domestic monopolies, contributing to "internationalizing" the region; its implementation would further impoverish the region's already poor human inhabitants by threatening traditional livelihoods; and the "risk contracts" posed a threat to the ecological balance of the region (Movimento em Defesa da Amazônia—São Paulo 1979; Movimento de Defesa do Meio Ambiente do Acre 1979). Ironically, subsequent accusations of "internationalizing" the Amazon would be directed against environmentalists themselves (Hurrell 1992).

The "risk contracts" campaign was not only, or even primarily, an environmental mobilization. It protested the military's expansionist and predatory development model for the region, as well as the closed system of decision making that had produced it—by implication, the authoritarian regime. Anticipating what would become the predominant discourse a decade later, the São Paulo document argued that an appropriate policy for using the region's resources "could grow out of the experiences of local populations, who show that it is possible to occupy the region without damaging its ecosystem. Yet, such a policy process requires two things incompatible with this dictatorship. Those are free development of research and scientific debate . . . and broad, open, democratic participation by all the people, who should have the final say in how best to take advantage of Amazônia's potential" (Movimento de Defesa da Amazonia, São Paulo 1979, 3). In summary, the Movement in Defense of the Amazon was part and parcel of the contemporaneous movement for political and economic transformation. To the best of our knowledge, it was the only time an antimilitary movement took on the environmental mantel in South America, although such articulations were common in the former Soviet bloc (Berg 2000; Darst 1997).

The campaign drew sustenance from multiple discourses, joining nationalism and the demand for democracy to environmental policy reform. Thus it attracted other opposition figures who saw environmental mobilization as a relatively safe way to contest the military government.[6] These included elite organizations of lawyers, journalists, and scientists as well as groups of labor unions, housewives, students, writers, and others. While strikes by São Paulo's metalworkers were met with troop deployments, the regime could discount environmentalist concerns as romantic and harmless. Nonetheless, the military government did crack down harder on this mobilization than it had on earlier ones, jailing five demonstrators at a protest march in Manaus in January 1979 and canceling the Fifth International Symposium of the Association of Tropical Biology, scheduled for Manaus in February. Subsequently the government moved discussion of its new plan for timber harvests into a long series of closed Inter-Ministerial committees and debates, demobilizing the protesters. Some individuals continued their environmental activism (interview with Cunha 1991), but the larger movement faded. Opposition political leaders shifted their focus to other concerns, leaving their former environmentalist allies feeling abandoned.

This hybridity of the campaign was simultaneously its strength and its weakness. Although the Movement in Defense of the Amazon served to publicize the issue, winning media coverage and a variety of allies among opposition organizations, it neither expanded opportunities for citizens' participation in policy making on the Amazon nor consolidated a national base for the environmental movement. Nor, in fact, did it generate anything like a serious debate among environmentalists on what, under different circumstances, might be the best way to regulate economic activities in the Amazon region. Instead, in the intensely politicized atmosphere of the time the question was framed as one of nationalism, privatization, and democracy, all of which helped to build bridges between environmentalism and other parts of the opposition to the military regime. Left out of the campaign was the problem on the ground of what to do about logging.

Building Local Networks: Rubber Tappers and Their Allies

The story of the Acre rubber tappers[7] and their struggle to prevent encroaching cattle ranchers and land speculators from wiping out their traditional extractive activities has been told many times (Keck 1995; Keck 1998; Keck and Sikkink 1998a; Mendes 1989). Resistance by rubber tappers in the state of Acre during the mid-1970s quickly built a network of domestic allies, both in Acre itself (with opposition politicians, the Catholic Church, and other social movements) and nationally (with the National Confederation of Agricultural Workers, or CONTAG; the Central Workers' Union, or CUT; and the Workers' Party, or PT). Acre had become a state in 1962, giving it a longer institutional history and more robust social networks than existed in neighboring areas like Rondônia, which had been a sparsely populated federal territory until 1981 (Keck 1998).

Despite the enormous difficulties and power asymmetries that the rubber tappers confronted, it was nonetheless a good time to be organizing.[8] In an associational sense, Brazilian civil society was awakening from a long sleep, and a variety of social movements organized in Acre in close proximity to the Catholic Church. In 1975 CONTAG sent an organizer to Acre, who worked closely with the rubber tappers and brought lawsuits on their behalf. They participated in the resurgence of the opposition MDB in the legislative elections of 1978 in Acre, and in organizing the Workers' Party and the CUT. The hardships for organizers of the rubber tappers were daunting, not the least of which was that for much of the year the only way to get from one part of the state to another was by water. Nonetheless, organize they did. Using the courts at the same time as they engaged in direct action to stop the tree cutters, they mounted an effective resistance to further encroachment into the forest, and began to plan for locally sustainable alternatives. In so doing they both reduced the speed of deforestation in the state and attracted the attention of sympathetic outsiders. They already had the attention of landowners. In 1981 the murder of Wilson Pinheiros, president of the rural workers union of Brasileia, Acre, brought leaders of the new PT (including its president, Lula, elected national president in 2003) and of the CONTAG to Acre for a protest rally. The core group of rubber tapper organizers and their supporters in Acre was also the group that was in the process of organizing the PT there. After the rally, five people—both national and local leaders—were indicted under the National Security Law for inciting to riot, guaranteeing their continued association (Keck 1995).

The rubber tappers' situation also attracted the attention of scholars (Brazilian and foreign), and at least two of those who began Ph.D. research in the region ended up actively engaged in the political process. One of them was Tony Gross, whom we met in chapter 3 as a coordinator of the Brazilian NGO Forum preparing for the UN Conference on Environment and Development of 1992. Having arrived in the region as a young political scientist from Oxford doing fieldwork, he gradually began to build international support for the rubber tappers, working with the British NGO Oxfam. Susanna Hecht, a professor of geography at the School of Architecture and Planning at UCLA, who had been following the incursions of the cattle frontier, became a fierce advocate, as did many of her students, who subsequently went to the region. Like Gross, Mary Allegretti, an anthropologist from Paraná, abandoned her dissertation and began to collaborate with Mendes and the others; she was soon working with them to devise practical solutions to problems ranging from educational and health services to new forms of land tenure.

This collaboration developed the idea of extractive reserves—conservation units in which traditional extractive populations had the right to remain on the land and continue to engage in extractive activities such as rubber tapping and nut gathering, and where measures would be taken to safeguard these activities and make them economically viable (Hall 1997). The extractive reserve embodies a classic enabling strategy in that both state and societal actors need to collaborate continually to achieve the reserves' ends. State actors are needed to formally establish both the overarching legal framework for collectively held property and specific territorial reserves. In the Amazon they also need to help protect and maintain those land grants.[9] And everyday activities by resident populations are critical for providing the intended conservation and livelihood outcomes of the reserves. Although the economic viability proposal was always rather tenuous—without subsidies for environmental services it remains hard to imagine that traditional extraction will ever produce enough revenue for economic security—it was nonetheless attractive to many outsiders. As a proposal originating from the region itself, it offered hope that something might actually be done to stem the destruction.

Transnational Influence on Preserving the Amazon

An explosion of environmentalist attention to the Amazon at the end of the 1980s responded to international as well as domestic political opportunities. A coalition of mostly North American environmentalists and Brazilian forest dwellers broke through the seemingly intractable refusal of governmental and private actors in the region to deal with the implications of deforestation. Oddly enough, the usual roster of active Brazilian environmentalists was not a significant presence. Instead, at the center of the struggle were groups with which only a few Brazilian environmental organizations had established contact to that point: rubber tappers and indigenous peoples on the western frontier. The avenues through which these struggles became known were unexpected as well: the World Bank and the Inter-American Development Bank (IDB), both in Washington, and the U.S. Congress.

The advent of civilian rule in 1985 did not win greater influence for environmentalists in federal policy for the Amazon region. The military retained control over key aspects of Amazon policy, still defined in national security and economic terms (Zirker and Henberg 1994). It remained at the center of the highly secretive Calha Norte project, intended to increase security in the border regions, and the development of SIVAM, a satellite imaging system intended to serve simultaneously the needs of air traffic control, military defense, interdiction of drug and timber smuggling, and environmental monitoring (Costa 2001). In his study of geopolitics in the region, Foresta even suggests that the Amazon was used as a safety valve for containing military dissatisfaction with the negotiated transition to civilian rule: "The remoteness of the northern Amazon frontier from the national ecumene, and its relative unimportance to national welfare, made it an ideal place to give vent to geopolitical thinking, acting as a sink for military energies that might otherwise go into political misadventures" (Foresta 1992, 139). When the centralized federal bureaucracy yielded political control over the region to elected state governors and legislators, local politicians depended on (dwindling) federal transfers and development projects for patronage resources with which to guarantee their political futures. This made them unreceptive, to say the least, to environmentalist efforts to curtail those very projects.

Members of the national environmental network paid little attention to the Amazon in the mid-1980s as well. Although they fought successfully for constitutional language designating the Amazon as a national patrimony and requiring that its environment be preserved, for most it was only one ecosystem among others on a list, like the Pantanal and the Atlantic Forest (Hochstetler 1997). Very few environmentalists from southern Brazil, the heart of the movement, had ever been to the Amazon. Environmentalists thus missed the boat on the more substantive legislative debate over the reintroduction of the Forest Plan of 1978, now modified to include economic and ecological zoning of the region. The revised plan was introduced in

Congress as one of the military government's last pieces of legislation in February 1985. The Amazonian states did as much to block this legislation as environmentalists had done earlier, objecting to federal meddling in what they saw as a state domain. The next moment of mobilization over the Brazilian Amazon largely bypassed both sets of actors.

The Washington connection was brokered by Gross and Allegretti, who became ambassadors for the rubber tappers' struggle. They were responsible for bringing it to the attention of the NGO activists in Washington, who were in the early stages of a campaign to challenge the environmental impact of multilateral development bank loans; British ecological activists associated with the magazine the *Ecologist*; and others. As for the multilateral bank campaign, it became an important source of leverage over Brazilian government policy in the region. The Washington campaigners were already protesting the impact of the World Bank's Polonoroeste project in Rondônia, next door to Acre, which combined a road-paving program with efforts to rationalize the chaotic settlement process in the state and protect ecological and indigenous reserves; instead, the project merely facilitated movement further into the forest. In 1985 the campaigners succeeded in getting disbursements temporarily suspended because the government was ignoring the environmental and indigenous components of the loan agreement.

Sensitized to the impact of road building in Amazônia, the Washington campaigners had begun to look at an IDB road project in Acre, about which they knew very little. Contact with the rubber tappers and their network of allies gave the campaigners access not only to information but also to vastly increased legitimacy for their efforts. They brought Mendes to Washington and Miami, where he won support for the extractive reserves proposal from both the World Bank and the IDB (Bramble and Porter 1992; Keck 1995). In addition, the campaign persuaded the U.S. Congress to pressure the World Bank to make reforms both internally and in implementing the Polonoroeste project. Later the bank made local participation a more central part of its Planafloro project in Rondônia, but the state's history was quite different from Acre's: local organizations were much weaker, and the state's politicians were able to ignore external pressure for much longer (Keck 1998).

It is worth remembering that by the time Chico Mendes and Mary

Allegretti persuaded the First National Congress of Rubber Tappers in 1985 to frame their struggle for land rights in the language of preventing deforestation, the Acre group had more than a decade of fierce struggle under its belt. At that same meeting it formed the National Council of Rubber Tappers (CNS) to support the proposal for extractive reserves (Grzybowski 1989, 15; Schwartzman 1989). The proposal also won support from the Brazilian special secretary of the environment, Paulo Nogueira Neto. On 30 July 1987, in Portaria 627, the Ministry of Agrarian Reform and Development created a legal instrument for establishing extractive reserves (Schwartzman 1989, 152). That same month Chico Mendes received the Global 500 award from the UN Environment Program, presented to grassroots activists, environmental organizations, and leading public figures for their contribution to environmental protection. The cattle ranchers, undeterred, remained determined to push further into Acre's forests, and since Chico Mendes was the central organizer of the resistance to their advance, they had long been trying to kill him. Finally, they succeeded.

Despite all this recognition among a specialized public, Mendes's killers had every reason to expect impunity when they shot him in his home on 22 December 1988. The international outcry over Mendes's murder stunned Brazilian authorities and Brazilians generally, most of whom had never heard of him. The scale of the international response included a front page story in the *New York Times* (24 December 1988). Domestically, it led to one of the first successful prosecutions of landowners in the region. More broadly, Mendes's death activated the diverse communities and individuals whose paths had crossed his over the previous decade and promoted the discourse of socio-environmentalism in both new and old organizations in Brazil and abroad.

After the Boomerang

In the decade following Mendes's murder his collaborators—both domestic and international—transformed environmental policy for the Amazon. They did this not by changing the laws, stopping the advance of the agricultural frontier, or prevailing on those who controlled more centrally located political institutions to think green, but through consistent and dedicated actions that relied on partnerships where they could find them and drew upon social capital from a variety of networks. They also built networks of their own, and with them expanded the reach of environmental information and campaigns.

The period prior to Mendes's death saw a mustering of alliances among civil society organizations in an attempt to pressure the state to stop predatory expansion; both in Acre and in Rondônia (see Keck 1998) they used what local channels were available and whatever external leverage transnational alliances afforded them. Although there was always some degree of collaboration on the part of individuals in the state (in the form of leaked information, rides from one part of the state to another, use of equipment, and the like), this was an essentially adversarial period. The boomerang itself is an adversarial strategy, which governments may resent for the international pressure and loss of sovereign control that it brings (Hochstetler 2003). The second half of the 1990s opened up many new opportunities to influence policy in the Amazon. While adversarial politics did not disappear, its predominance gave way to a much greater degree of collaboration, especially at the federal level. Both state and societal organizations grew stronger and more interdependent during these years. At the same time, relations between the World Bank and Amazonian NGOs, though still mistrustful, lost much of their combative edge. So did the relations between the Brazilian national state and foreign governments (Kolk 1998).

Specifically, the impact of new information on high rates of deforestation, intensified by the shock of Mendes's murder and the international reaction to it, set off a chain of events that brought about unforeseen levels of collaboration between the Brazilian state, local activists in the Amazon and their national and international allies, multilateral development banks, and foreign governments. The presence of new funding for environmental protection was a crucial part of this process, especially for the Brazilian government (Kolk 1998; see also Hochstetler 2002a). While increased conservation dollars tracked many American donors' interests in global environmental concerns such as biodiversity (Lewis 2003), more socially oriented groups like the rubber tappers also maintained significant international support for their livelihood concerns (Pizzi 1995), often from European sources. The Pilot Plan for the Amazon, funded by the G-7 and discussed further below, funded both kinds of activities, exemplified by its support for extractive reserves.

The Advances of Socio-Environmentalism— Into the State

The publicity associated with the death of Chico Mendes strengthened the political resolve of local activists by validating the importance of their struggle. Between 1989 and 1992 forest peoples held a long string of meetings—building alliances and other fleeting organizational forms—among groups traditionally at odds, such as rubber tappers and indigenous populations, and even small farmers and the landless movement. These were not always frictionless, and sometimes they broke down altogether or led nowhere. Nonetheless, these served as sites for sharing experiences, recognizing presence, attracting publicity, and eventually attracting funding as well. Although under ideal circumstances it might have made sense to proceed more slowly, by the beginning of the 1990s there was a rush to establish new conservation units to safeguard traditional populations where possible, given that a new opportunity might not present itself so soon. Unlike environmentalists more generally, Mary Allegretti and rubber tapper leaders maintained good relations with the environmental secretary, Lutzenberger. A Center for Support to Traditional Peoples was set up in IBAMA in 1990 to work with forest dwellers in establishing extractive reserves. This gave the followers of Chico Mendes their first real foothold in the federal government structure, and they used it to further build the capacity of social movement organizations in the region.

At this time the extractive reserves proposal evolved from an experimental undertaking to a recognized form of conservation unit, soon included at the urging of environmentalists in such undertakings as the Rondônia Natural Resources Management Project, funded by the World Bank; in this project it was little more than window dressing for a state government more interested in extracting resources than managing them, despite NGO pressures (Keck 1998; Rodrigues 2004). All these efforts required support and political protection from the outside. This assistance appeared to be forthcoming, as the rubber tappers and indigenous peoples of the Amazon (along with other iconic groups in India, Indonesia, and elsewhere) fired up an international conversation about the link between conservation and sustainable livelihoods that had in principle been part of the "sustainable development" mantra from the beginning.

Support inside Brazil was also crucial. Mary Allegretti, who had collaborated with Chico Mendes in creating the Rubber Tappers Project (Projeto Seringueiro) in Acre, forerunner of the extractive reserves proposal, promoted the idea of extractive reserves tirelessly both inside and outside Brazil. At the same time, nongovernmental organizations created projects and undertook study after study designed to imagine ways that extractive reserves could be economically viable. The problem was the definition of economic viability, and study results were contradictory. Bee keeping, processing Brazil nuts, manufacture of sandals with natural latex, and a variety of agro-forestry plans were floated, and many tried. This was the period when Ben and Jerry's came up with Rainforest Crunch and the Body Shop gave out folders about extractivism as they sold natural cosmetics.

Allegretti, a native of the far southern state of Rio Grande do Sul, arrived in the Acre rubber region in 1978 to do fieldwork for her degree at the University of Brasília. She first met Mendes in 1981, when he was leader of the rural workers union of Xapuri. She worked with him on Project Seringueiro, with support from CEDI, on education and health programs and to reinforce the rubber tappers' organization against the advance of the ranching frontier, and on the organization of the first national rubber tappers' meeting in 1985. She lobbied hard to transform the extractive reserves project into a national policy, watching it be incorporated in 1987 into the National Agrarian Reform Policy and two years later into the National Environmental Policy. She was a consultant to the Indigenous Peoples' Division of the IDB in 1994 and secretary of planning of Amapá from 1995 to 1996, where she helped institute the first state-level sustainable development plan. She began a run for vice-president on the Green Party slate led by Alfredo Sirkis for president in 2001, but dropped off the slate because of differences over how the campaign was being managed.

The experience in Amapá led to Allegretti's appointment as secretary for Amazônia in the Ministry of the Environment when it was headed by José Sarney Filho (Sarneyzinho) in 1999, during the second government of Fernando Henrique Cardoso. The son of a powerful senator (and former president), José Sarney, the new minister was sensitive to environmental concerns and had been a member of the Green Front during the Constituent Assembly. After leaving the ministry he joined the Green Party. In office he gave Allegretti considerable support in seeking solutions for the continuing deforestation in the Amazon region. She was an activist secretary, not only in the area of demarcating extractive reserves but also in creating new instruments for implementing existing legislation, such as the Integrated System for Monitoring, Licensing, and Control of Deforestation. This was an important move, as it enabled the licensing agency to distinguish between legal and illegal deforestation. According to the Forest Code, landowners in the Amazon were required to leave 80 percent of their landholding untouched, a stipulation that landowners opposed and routinely violated. The new system required that a baseline satellite image of the property be provided along with each application for a license, making it possible to use future imaging to detect violations.

In addition to efforts to enforce existing laws, Allegretti began a difficult process of negotiating with state governors over programs and projects that could benefit the environment but that the governors believed would benefit them politically as well. The governor of Pará had obstructed Allegretti's congressional confirmation, making her path more difficult than that of any of her colleagues (*Folha do Meio Ambiente*, July 1999, 18). Negotiating with the governors meant giving priority to activities like ecotourism development and other projects which might not have had the greatest impact in terms of preserving the forest but had the potential to build more trust between the state governments and the ministry. This initiative was especially positive because it was the first direct acknowledgment of the obvious fact that although the federal government had a great deal of responsibility for what went on in the Amazon region, it had almost no control over it.

There was progress. Along with new licensing and monitoring procedures came a decision to release satellite pictures of the Amazon before the burning season began, in order to determine where to concentrate the greatest efforts. New conservation units of various kinds were created, including extractive reserves, national forests, and areas of indirect use (to be used only for science and research). Table 4.1 shows that 1998–2005 was Brazil's most intense period of creating Amazonian conservation units ever. After negotiation, the WWF proposal for a program to place 10 percent of the Amazon Forest under a strict protection regime was transformed by 2002 into a program that applied resources to both strict protection and sustainable use areas. As other parts of the Cardoso and Lula administrations pushed hard to get new road projects approved for the region, the

Table 4.1 Federal Conservation Units in Amazônia

DATE OF CREATION	NUMBER OF CONSERVATION UNITS CREATED	NUMBER OF HECTARES PLACED IN CONSERVATION REGIME
1959–1984	26	12,267,635
1985–1989	22	12,665,535
1990–1993	18	6,720,472
1994–1997	1	250,190
1998–2002	34	13,823,654
2003–2005	19	8,682,722
2006 (to April 15)	7	6,301,486
Total	127	60,711,694

Source: Ministério do Meio Ambiente, Secretaria de Biodiversidade e Florestas.

secretariat for Amazônia tried to increase the number of conservation units along the edges of the highways. However the secretariat for Amazônia was greatly weakened during the Lula government, after Marina Silva (senator from Acre, discussed below) became minister of the environment; this weakening eventually led Allegretti to resign from her position, after which its size was substantially reduced.

New Collaborative Environmental Networking

Besides greater space for translating concerns with the Amazon into state policy, there were advances in the size and sophistication of Brazilian NGO networks working in the region. This simultaneous reinforcement of both state and civil society actors on the environment was an important element in the move from adversarial (blocking) to collaborative (enabling) relations. For example, the Brazilian Network on Multilateral Development Institutions (Rede Brasil), represented a deliberate transfer of a political technology from actors in the multilateral bank campaign to Brazilian NGOs. Steve Schwartzman, an anthropologist working at the Environmental Defense Fund (now Environmental Defense) who had become involved with the

Amazon through his dissertation research among the Parará Indians in the Xingú National Park, had been the main point person on Brazil in Washington for the multilateral bank campaign. He was one of the people whom Allegretti and Gross went to see in 1985 and one of those through whom the foreign campaigners established a relationship with the Acre rubber tappers. His continuous collaboration with Brazilian colleagues made Schwartzman better aware than most of the inherent tensions in a relationship that had Brazilians providing information for campaigns and activists in Washington mediating relations with the bank, their activities ranging from the collection of documents and the maintenance of archives to direct conversation and sometimes confrontation with bank officials.

Schwartzman therefore resolved to teach a group of Brazilian activists what he and others had learned about how the multilateral development institutions worked, and the strategies that the campaigners had employed. Schwartzman secured funding from the MacArthur Foundation to hold a meeting in Brasília, out of which was formed a network to monitor the activities and plans of the World Bank and other multilateral development banks in their country. In one of its first major actions, the Rede Brasil persuaded allies in the Brazilian Congress to mandate that the planning ministry pass on to them a copy of the bank's confidential Country Assistance Strategy, thus transforming the document into one for public discussion (Vianna 1998). The Rede Brasil leader Aurélio Vianna remarked in an interview in 1998 that Schwartzman had shown remarkable grace in knowing when it was time to walk away and leave the Brazilian network to set its own path (interview with Vianna 1998).

A much larger network of organizations in the Amazon Working Group (Grupo de Trabalho Amazônico), also begun with some external support, was a powerful stimulus to reconfiguring NGOs in the Amazon region. It began in 1992 when a small group of Amazonian NGOs met in Porto Velho to discuss the opportunities being opened up by a very large project then being designed as a result of a decision by the G-7 summit in 1990. Begun in 1992, the Pilot Program for Conservation of the Brazilian Rainforest (PPG-7) is administered by the World Bank but run out of the Brazilian Ministry of the Environment. Major donors are the G-7 countries, the Netherlands, and the European Union. Conceived from the beginning as including local communities, the program has advanced the most with regard to extractive reserves and in the area of demonstration projects, the great majority of which were proposed and carried out by community and nongovernmental organizations. The PPG-7 is exceptional among multilateral projects in that it has specifically financed capacity building among NGOs and community organizations in a separate component of the project.

Today the Amazon Working Group network claims 602 organizations as members and is structured in all nine states of Amazônia. True to the socio-environmentalist vision, it includes NGOs and social movements representing environmentalists, rubber tappers, babaçu coco palm workers, Brazil nut pickers, river dwellers and fishermen, family farmers, young people, women, human rights groups, technical assistance groups, and others.[10] Activities related to the pilot program helped to promote collaboration between government agencies and community organizations in the 1990s. They also contributed to a pattern of growth by project prospecting among Amazonian NGOs.

In part because of the impetus of this sort of external support, many new environmental organizations were also founded in the 1990s. Only a few long-established groups had existed in the Amazon before the 1990s, and the number remains small. In 2002 the IBGE counted 101 organizations dedicated to "environment and protection of animals" in the North of Brazil, employing 604 people (almost all in Rondônia, Amazonas, and Pará). This represents just 6.35 percent of the organizations in the country, although the figure is surprisingly high given the North's 2.54 percent of nonprofit organizations overall (IBGE 2004, tables 17, 19). Without a doubt, some of these new organizations are what fellow activists would call "opportunists." Nonetheless, many have serious conservation and socio-environmental aims.

New Environmental Organizing: The Fundação Vitória Amazônica

One example of the new kind of organization formed after Mendes's murder is the Fundação Vitória Amazônica (FVA), created in 1990. Conscious of the growing national and international attention to the region, the professionals in Amazonas state who founded the group wanted to create a strong regional presence among environmental organizations. The group had close relations with regional research institutes—its founder Muriel Saragoussi had worked at the Institute for Amazon Studies (INPA) for eleven years, and the foundation's council included university professors and INPA researchers. The foundation began with an initial patrimony from a series of paintings by the painter Sepp Baendereck[11] and other local donations; it quickly developed close ties to Conservation International and the World Wildlife Fund. One of its early directors, Carlos Miller, had worked for WWF and CI in Washington before going to work for Vitória Amazônica (interview with Miller 1991).

The group decided to focus on making conservation viable in the Parque Nacional de Jaú, a national park in the Rio Negro basin that had been created ten years earlier but remained a park on paper only. They did a census of the thousand people living in the park and developed a co-management contract with IBAMA, leading to an agreement on technical and scientific cooperation. Between 1991 and 1996 the group worked with communities in the park to produce a management plan.[12] Working with the park's inhabitants was a commitment of the group from the beginning, as they were convinced that they could not protect the park without the collaboration of those who used its resources. They understood that asking people to use the resources less required that they be compensated, or that alternative livelihood activities be developed. Most adults living in the park were illiterate and lacked civil documents like birth certificates. The FVA helped to secure documents, promote literacy and health care, bring electricity to homes through solar energy (collaborating with the University of Amazonas, Eletrobrás, and Centrais Elétricas de Amazonas), and promoted artisan activity using forest products. It also worked in towns directly outside, training teachers with methods devised earlier in Acre and developing a citizen education program. This work was somewhat risky. According to Saragoussi, the foundation's work in Novo Airão had a political impact, producing a renovation of the rural workers' union and the formation of several associations through which citizens began to oppose domination by local elites.

FVA's networking experiences in the early 1990s showed some of the continuing difficulties in bridging the many regional and political divides in Brazil at the time. Because of the political impact of its activities, FVA had almost no partnership relations with the state government in Amazonas. Rather, its actions involved collaboration at the local, national, and international levels. FVA worked with a few other local organizations in initial meetings with representatives of the G-7 countries as the PPG-7 was being developed. Thousands of miles south of their meetings, social movements and NGOs meeting in the Brazilian NGO Forum for UNCED (see chapter 3) criticized their efforts, saying they had never heard of these upstart conservation groups and suspecting them of having been formed to illegitimately garner foreign funds (*Gazeta Mercantil*, 16 July 1991). The Society for the Preservation of the Natural and Cultural Resources of the Amazon (SOPREM), which had conducted environmental education in the Amazon since 1968 and was part of the meetings, shot back with a letter withdrawing from the forum, which it accused of behavior smacking of internal colonialism (SOPREM letter 1991). Forum members really should have known the work of SOPREM, but Miller admitted that the Fundação had an unusual profile at the time, being extremely well known internationally but less so nationally and locally (interview with Miller 1991).

With time, and in large part through the regular meetings of the Brazilian NGO Forum, these intra-Brazil differences were bridged. Saragoussi was even elected to represent the forum at some international meetings. Eventually FVA was part of the SOS Forests Campaign, which lobbied Congress to oppose the landowners' proposed forest code, was part of the Amazon Working Group, represented the group in the federal commission to manage the Central Corridor of Amazônia, and represented the forum in the national Biodiversity Foundation (while also serving as the alternate in another). Its members collaborated in writing the draft law and enabling decrees for the National System of Conservation Units, and a draft law on access to genetic resources that regulates part of the Biodiversity Convention. The group was also a member of the municipal environmental council in Manaus, debated socio-environmentalism in meetings of the PPG-7, and spoke for Amazonian NGOs in a number of other meetings (Lena 2002). Saragoussi replaced Mary Allegretti as secretary for the Amazon when she left the environment ministry in 2004.

The Continuing Confrontation of Environmentalism and Developmentalism

At the same time as there were more committed personnel prepared to work hard and more potential collaboration between government organs and societal organizations, other, stronger forces were undermining many of these initiatives. The economic crises of the 1990s led Brazil to accept harsh conditions for IMF bridge loans, complicating efforts to develop rational spending plans. In practice, this meant that until the government was certain of the availability of money required for the mandatory primary surplus, budgetary funds were not released for the purposes to which they were committed.[13] Project money was not disbursed, and although personnel were usually paid, new personnel could not be hired. Then late in the year, after the surplus had been secured, the remaining money was often released all at once, with the requirement that it be spent before the fiscal year was over. These financial boom and bust cycles wrought administrative havoc not only in the environmental organs of the state but overall.

Similarly, a commitment to run budgetary and balance-of-payments surpluses meant that exports were essential. The most dynamic area of export expansion in Brazil was the agricultural frontier, and soybean production spilled over from the savannah regions where it had initially spread and began to encroach onto the forest. At the same time, timber extraction, both legal and illegal, joined cattle ranching as an important motor of deforestation, while cattle ranching grew more profitable. In the face of an insistence on development at any cost, Greenpeace came to the region to fight back hard. The equivalent in the 1990s of the dramatic actions in Acre in the 1970s, when rubber tappers confronted plantation hands who were wielding chainsaws, was the picture of Greenpeace activists creeping up on illegal logging operations and splashing luminous paint on logs, which they would then identify in ports. These kinds of activities demonstrated a renewed opposition between environment and development in the region, against the hopes that the two concerns could be joined in socio-environmentalism.

Environmental Resistance: Greenpeace in the Amazon

Greenpeace established branches in several Latin American countries only in the early 1990s, when Tani Adams, a binational of the United States and Guatemala and a board member of Greenpeace-U.S., accepted a request from Greenpeace International that she become a regional organizer. Over the next few years she held meetings with environmentalists all over Latin America. In Brazil she put together a group of long-time activists who shared a socio-environmental outlook and were willing to try a different form of environmental activism—highly professional, well paid and well funded, and on a world scale. In fact, Adams's insistence that the new Greenpeace activists come from the socio-environmentalist camp eventually put her at odds with the British and German branches most influential in the international organization, and (together with other sources of conflict) caused her to resign from her organizing role. Even after Chico Mendes, not all foreign environmentalists were ready to accept socio-environmentalism's claim that the livelihood concerns of local inhabitants were critical for environmental protection, and some preferred a return to the "fortress conservation" that kept humans out of conservation areas.[14]

Even before Greenpeace had been formally launched in Brazil, it had greater name recognition than any other environmental organization. Moreover, it was bankrolled by Greenpeace International during its first years of operation, giving it an enormous advantage over other groups, who had to rely on their own fundraising. In the Amazon, Greenpeace resolved to confront the illegal harvesting of tropical timber, taking advantage of its multiple national platforms to coordinate campaigns among buyers and sellers and collect information on trans-boundary activity. The Amazon campaign's first coordinator, José Augusto Pádua, was an ecological campaigner from Rio de Janeiro who had become active in the movement in the 1970s and was highly respected in Brazilian environmental circles. In one successful campaign of the 1990s, Greenpeace joined with Friends of the Earth to have mahogany listed in annex II of the Convention on International Trade in Endangered Species (CITES).[15]

By 1999 Greenpeace Brazil had persuaded Greenpeace International to make the Amazon campaign an international priority, and the executive director of Greenpeace International, Thilo Bode, kicked off the Campaign in Rio de Janeiro on 31 May 1999. By 2000–2001 Greenpeace had become singularly effective in tracing illegal mahogany. In July 2000, for example, the campaign chief, Paulo Adário, led a four-month trip up the Xingú River looking for evidence of illegal harvesting. Finding a stack of illegally harvested mahogany logs alongside the river on Kayapó lands, the group dabbed ultraviolet paint on them and was able to track them to the sawmills for which they were destined. In October 2001 Greenpeace released a report called "Partners in Mahogany Crime," which named much of Pará's mahogany mafia. In the wake of the report Brazil announced a mahogany moratorium, but it was extremely difficult to enforce. In the meantime Greenpeace headquarters became almost like a bunker surrounded by bodyguards, and campaigners like Adário slept in a different place every night (Symmes 2002). The campaign continued to be energetic with respect to logging, and it successfully supported the creation of two new extractive reserves in southern Pará in late 2004. In September 2005 Greenpeace launched its "Friendly City" program, a campaign enjoining cities to agree not to allow illegal lumber to be used in their construction projects. Ten days after the campaign began, it had already signed up Manaus, Rio de Janeiro, São Paulo, Vitória, and a host of smaller cities.

The Nationalist Backlash

The presence of Greenpeace, Friends of the Earth, and WWF in Brazil increased the profile of some environmental campaigners, but also gave more energy to the continuing nationalist backlash, which identified conservationists with a foreign invasion. The widely publicized tours of the Greenpeace boats *Amazon Guardian* in 2000 and *Arctic Sunrise* in 2001 have been cited by some as a provocation that encouraged the creation of a parliamentary inquiry commission on NGOs in 2001; in any case, Amazonian politicians then heightened their rhetoric about the perils of "internationalization." Any search on the Internet for material on the Amazon is fairly certain to uncover articles claiming that Brazilians are not allowed to enter certain parts of the Amazon, so great is the foreign grip on the region. Internationalized campaigns—for certification of forest products, for example—have little meaning to workers or extractors in the region, who need a viable form of livelihood. Although their feelings on the subject are fueled by the claims of powerful logging interests, many in southern Pará, for example, mistrust the intentions of outsiders—whether foreign or from the Brazilian government—even more. This is the region which has been a champion in violence against organizers, where Sister Dorothy Stang told a journalist in 2002 that she had received her most recent death threat three days before, "after helping disarm three pistoleiros trying to evict farmers from land claimed by a wealthy rancher." " 'If I get a stray bullet,' the sister said cheerily, 'we know exactly who did it' " (Symmes 2002). Pará offers one of the most graphic illustrations of the absence of government and the ceaseless contestation of the claims of socio-environmentalism—and disregard for its proponents.

The Return of State-Led Development

During the late 1990s the macroeconomic imperative of increased export earnings provided a hospitable environment for further encroachment into the forest. Logging firms, including some of those that had already decimated the forests of Borneo, came in to extract high-value timber, destroying much of what surrounded the tall mahogany trees in the process. Soybean production occupied ever more land. New road-building programs, undertaken in the second half of the 1990s, made ranching, agriculture, and logging viable in parts of central Amazônia that had previously been spared (Fearnside 2001; Margulis 2003; Greenpeace 1997; Soares Filho, Silveira, Nepstad, Curran, et al. 2005).

Many Brazilian socio-environmentalists hoped that the presidential administration of the long-time PT leader Luis Inácio Lula da Silva (Lula) could resolve many of these related problems after 2003—rural violence over land claims, predatory deforestation, Brazil's economic stagnation, and its persistent social and economic inequalities. The Amazon seemed like a good site to begin this struggle, especially when Lula made the "Acre Connection" central in his environmental administration (not surprisingly, given his own early connection with the Acre group). Mary Allegretti was the only secretary in the ministry who carried over from Cardoso's government to Lula's. With the appointment as environment minister of Marina Silva, the charismatic senator from Acre, Lula left most environmentalists overjoyed, convinced that her presence would boost the priority of the environment and particularly the Amazon on the government's agenda. Her retention of Allegretti meant that two of Chico Mendes's closest associates were now in the ministry in Brasília; another, Jorge Vianna, was governor of Acre. Still, some worried. Lúcio Flávio Pinto, the journalist from Pará discussed above, worried that the new minister's understanding of the Amazon was too narrowly grounded in Acre, and that she might be unprepared to deal with the complexities of the larger region. Others worried that she would not have the power to create transversal linkages among ministries with influence in the Amazon—something that others had tried to do before, and failed. Three years into the administration the doubters are winning, notwithstanding some achievements by the new minister and high personal regard for her and her story.

Marina Silva met Chico Mendes in 1978, when she was twenty. She grew up in a rubber-tapping area seventy kilometers from the capital of Acre, a long trip by river from anywhere. There were no roads, no schools, no clinics. At five she went with her family when it tried its luck in Belém but had to return, unable to make a living. Only in her teens did Silva gain access to the education she sought, when she moved to Rio Branco in search of medical treatment for hepatitis. Active in the ecclesial base communities of the Catholic Church in the 1970s, she considered becoming a nun, but found her vocation in activism, joining the student movement and neighborhood association. After receiving help from Dom Moacir Grecchi to go to São Paulo for medical treatment, she returned to Acre and entered the university. There she discovered Marxism and joined a semi-clandestine organization, the PRC. She earned a degree in history at the University of Acre in 1985 and began to teach there.

With Mendes, Silva founded the Acre branch of the CUT, and for the next two years she was vice-coordinator to his coordinator. In 1986 she ran for federal deputy on the PT ticket, and although she was one of the five candidates who received the most votes, the PT did not meet the threshold to win a seat. In 1988 she won the most votes of any candidate in the municipal council elections in Rio Branco, becoming the only councilor on the left. In a tumultuous move, she publicly returned various per diems, housing assistance, and other political perquisites, revealing on television and in the newspapers just how much municipal council members were earning. In 1990, as a candidate for state deputy, she was again the most-voted candidate. Two other state deputies were elected from the PT, and the party's gubernatorial candidate, Jorge Vianna, made it to the runoff election. However, after an activist start as a deputy, Silva saw her health fail again. This time more complete tests in São Paulo confirmed what she had long suspected: she was the victim of heavy-metal contamination, probably contracted while still living on the *seringal*, or rubber plantation. In 1994, under an extremely restrictive health regime, Silva was elected senator for Acre, and she quickly became a national and international figure as well as a regional one. Reelected in 2002, she instead took up the position of minister of the environment in Lula's government.

Despite their initial hopes, environmentalists quickly became disillusioned with Lula's administration because of its evident lack of a program of action on the environment. In March 2003, after discovery of areas of soy plantation that had been contaminated by genetically modified (GM) soybeans, Lula agreed to their legalization without opening the issue for broad discussion. The administration took an unexpectedly firm and positive stand on GM crops, reversing the position the PT had taken while in the opposition and infuriating environmentalists (Hochstetler forthcoming). Personal interventions by Lula's powerful chief of staff, José Dirceu, blocked an international agreement on labeling standards *and* ensured passage of a national law allowing GM crops despite Marina Silva's mustering of opposition to GM among social movements (interview with Lisboa 2005).

The administration also put road-building programs into its development plan, continuing on the potentially disastrous course set by the Cardoso administration's ambitious Avança Brasil (Brazil Advances) program for the region and promising to increase the accessibility—and thus probable deforestation—of some of the most remote areas (Laurance, Cochrane, Bergen, Fearnside, Delamônica, Barber, D'Angelo, and Fernandes 2001). One of the most problematic proposed roads is BR-163, which would go through the so-called Terra do Meio (Middle Land) in southern Para, at the heart of contestation over illegal logging, land grabbing, indigenous areas, rubber-tapping areas, and cattle ranches (Soares-Filho, Silveira, Nepstad, Curran, et al. 2005). The debate over BR-163 prompted an inter-ministerial working group to propose remedies, its actions closely monitored by Roberto Smereldi of Friends of the Earth, Greenpeace, and other environmentalists. However, despite proposals to implant conservation areas and demarcate existing indigenous reserves in the areas around the new roads, the government was reluctant to confront the concerted opposition to the proposals from powerful governors.

In October 2003, not even one year into his administration, more than five hundred NGOs and social movements sent Lula a letter criticizing his environmental policies in the strongest possible terms. Two of the topics that they singled out for special emphasis were the administration's support for GM crops and its endorsement of large infrastructure projects in the

Amazon. The letter made these social movements the first ones to publicly break with the administration over its policies, many of which had surprised and dismayed the PT's historic backers (Hochstetler 2004).

Well into Lula's first term in office, the state of the forests seemed to be getting worse, not better. Between August 2003 and August 2004, an area of 26,130 square kilometers was deforested, more than the year before, which had been the second-worst year since records began to be kept (Secretaria de Biodiversidade e Florestas 2004). Illegal logging and forest burning were still out of control. Nearly half the deforestation during the year was in the state of Mato Grosso, where soybeans for export—soon, possibly, to be genetically modified soybeans—were pushing back the agricultural frontier into the Amazon. Frustrated with stonewalling from other ministries, the primacy of partisan (PT) considerations over competence in the appointment of new officials in the ministry, and the weakening of the secretariat, Mary Allegretti quit in 2004, to be replaced by Muriel Saragoussi of Vitória Amazônica.

The deforestation rates gained international attention, but it was the murder of Dorothy Stang that transformed the possibilities for action overnight. Once again it was the combination of external scrutiny with outrage at internal lawlessness that provided the combustible mix. That Stang was a nun would have mobilized a portion of the Catholic Church; that she was an American nun, despite having taken Brazilian nationality long ago, made international headlines. The resulting pressure allowed the immediate demarcation of protected areas proposed for the region, as well as the elimination of some of the dubious instruments traditionally used by land grabbers to sustain their land claims. Even after her death, however, the national newspaper *Jornal do Brasil* noted the "bloody struggle" between environment and development that continued within the administration (*Jornal do Brasil*, 27 February 2005), and that still seems ineradicable in the Amazon region.

The challenge for the next period of the region's development is already, ironically, visible: it arises from the slow recovery of Brazil's economy, and from the determination of recent national presidents, from Fernando Henrique Cardoso to Lula da Silva, to take up regional planning again. As new programs appear ready to repeat the experience of the last phase of national developmentalism in the region, will new, giant projects simply steamroll the small-scale alternatives that have been carefully nurtured over the last decade? It seems more likely than not, and in their article in *Science*, leading Amazonian researchers pointed to growing signs of crisis in the forest, including the severity of its most recent drought (Laurence et al. 2005). However, this time around the developmentalists are more likely than in the past to meet with resistance in the region, and from those who have taken the region's struggles inside the state itself.

The Amazon is full of paradoxes. Communication networks and networks of roads demarcated the territory according to state plans for the region, connecting it to national and world markets and making it susceptible to control by security forces. New communication networks, linked to the Internet, facilitate both the further linkage to international commodity exchanges and the rapid mobilization of national and international resistance to predatory activities. The roads that brought floods of migrants to the region are, according to research by Brown and his colleagues (2005), the routes by which environmental ideas and progressive politics also make their way slowly into new areas. Increasing urbanization of the region favors less intensive forms of land use like cattle ranching, resulting in more deforestation "since the agents of deforestation shift from small farmers to larger urban-based entrepreneurs with enough capital to deforest large areas and to keep the forest from returning" (Moran and McCracken 2004, 35). At the same time, urban areas concentrate resources that facilitate organization. The highly secretive, military-run SIVAM program for satellite surveillance of the region, intended to protect borders and detect illegal activities, is also a monitor of environmental destruction and may provide an early warning system for new fronts of deforestation.

Just as it did externally, the story of Chico Mendes had an impact in the Brazilian politics of the Amazon that was much greater than one might have expected. However, the reasons for its persistence in domestic and in international politics were different. For foreign publics, the story seemed to make explicit a set of relationships involving poor people, extractive activities, ranchers, the police, and other state organs; it also suggested remedies: for example, that helping traditional extractors stay in the forest was a good way of helping the forest itself. The story was powerful because it was also hopeful—if Ben and Jerry's sold enough rainforest crunch, if the Body Shop sold enough creams and shampoos, and if conservation organizations could help rubber tappers, indigenous peoples, artisan fishermen and women, and other benign residents of the region to stay there, then perhaps there was hope for saving the rainforest in spite of the powerful forces that were arrayed against it.

Domestically, people knew that there was more to the story, that Acre was far from typical of the region, and that the reasons for the rubber tappers' success there, such as it was, had only partly to do with international support. Rubber tappers had built a web of elite allies in the state that their counterparts had done nowhere else largely because the organizing of rubber tappers in Acre had been a highly *politicized* process. Some of the organizers, reportedly including Mendes himself, belonged to clandestine political parties of the left during the dictatorship. At the same time, the rubber tappers had worked closely with the Catholic Archdiocese, with Catholic activists in other social movements in Rio Branco, with the student movement at the University of Acre, and with opposition leaders in the state in building the electoral force of the legal opposition MDB in the 1978 elections. Rubber tappers in Rondônia were weakly organized and highly dependent on clientelistic politicians and a paternalistic Catholic hierarchy. Impetus for organization in the state had to come from elsewhere, and was much more difficult. The configuration of actors in Pará and Amazonas has still other characteristics. It was because the struggle in Acre was already connected to the political and social networks of Acre state that it provided such a propitious starting point for government programs.

There were almost no regions in the Amazon rainforest where such a web of relations had already existed at the time the environment ministry began seeking out ways to find human support for forest preservation there. For government to *do* something, it must establish its authority over the territory; when attempting to do so, it is helpful to have allies. Untypical as it was, Acre became a pilot case for socio-environmental policies—policies attempting at the same time to preserve and improve livelihoods and maintain the forest standing. This process reinforced the pro-forest alliances within the state at the same time as they provided experimental programs for testing the economic viability of extractive reserves, and also reinforced the rule of law. Governor Jorge Vianna is in his second term as head of what in Acre is called the "forest government."

It has been extremely difficult to find similar synergies in other parts of the Amazon, especially in Pará, the current (and long-standing) focus of conflict. International human rights organizations have mounted campaigns against widespread torture in the region, the continuing practice of slavery in remote areas, and the steady stream of violence against human rights workers. There are significant risks in supporting the establishment of governmental authority in the face of what amounts to warlordism. Until the Brazilian state is able—and willing—to bring enough coercive force to bear on the region to eliminate the worst of the purveyors of violence and establish the rule of law, it is hard to expect that new licensing procedures or forest principles will "stick." Economic incentives for reaping gains from long-term stewardship cannot compete with the economic incentives for rapacity, as long as impunity reigns. Nonetheless, the web of relations has been growing, responding to incentives provided by demonstration projects, new local organizing, new programs, and extraordinarily dedicated people who continue to risk their lives to change things.

The focus of people like Mary Allegretti when she was in charge of the Secretariat for the Amazon, as well as a part of the NGOs campaigning on forest issues, has always been on the peoples of the forest, in the belief that efforts to conserve forests without their collaboration would inevitably fail. Many foreign (and some domestic) conservationists have been frustrated with this approach, arguing that total protection was more reliable, and that if conservation had to depend on the peoples of the region, the forest would be gone before there was time to convince anyone that it was worth saving. In this view, what matters is results, and not whether the process through which the results were reached was democratic or participatory. As new, large "integral" conservation areas were demarcated in the Terra do Meio of southern Pará after the murder of Dorothy Strang, these questions will be asked again. But if multilevel governance is to slow the course of forest destruction, there must be local support—and for that to emerge, someone must guarantee that those on the side of the law are not simply setting themselves up for a bullet. In other words, environmental protection in this part of the world at this time—as in all of the last thirty years—really seems to require building a credible and responsive state capable of mediating among diverse needs and interests, enforcing laws, and protecting the lives of those who contest the continued predation of the region's forests and its people alike. Nothing less will serve. That such a scenario remains a utopian imagining for much of the region helps to explain why deforestation continues apace.

5

From Pollution Control to Sustainable Cities

Brazilian urban areas present both administrators and residents with major challenges. They suffer from too many people in too little space, inadequate infrastructure and planning of all kinds, and widespread violence and poverty (Bezerra and Fernandes eds. 2000; Caldeira 2000). At the same time, Brazilian cities are economic and cultural centers of national life. They continue to attract new residents who find opportunities there that are not available elsewhere. As of the 2000 census, 81.25 percent of Brazilians lived in urban areas, as did most Brazilian environmentalists. Their intimate relationship to their concrete habitats has encouraged the intrepid to work to transform their surroundings. In the mid-1990s 39 percent of all Brazilian governmental and nongovernmental environmental institutions focused on urban environmental issues (Crespo and Carneiro 1996, xxi).

The role of the national government is less central for these issues, as the Brazilian federal system gives much of the responsibility for controlling pollution to government at the state and municipal levels.[1] Urban environmental actors also work with fewer international allies and attract much less international attention than their "green" counterparts do. International development agencies have provided substantial resources over the years for sanitation and other urban pollution projects, for example, but their funding rarely attracts prominent headlines of the kind that appear over stories about combating deforestation. (The Cubatão case below is that rare exception.) A few global discourses have become important in urban environmental politics in Brazil, with global climate change and environmental justice introduced as new justifications for environmental action.

Our focus in this chapter shifts from Brazil as a whole to São Paulo state, reflecting the change in relevant political levels. São Paulo merits attention not because it is a representative case of urban environmental politics in Brazil, but rather because it is so extraordinary. Metropolitan São Paulo exemplifies the dangers, difficulties, and attractions of Brazil's large cities. *São Paulo não pode parar!* (São Paulo cannot stop!) was its motto from the 1950s until the 1990s, when pessimism finally took over (Caldeira 2000, 40–41). A population of around seventeen million in the metropolitan area makes São Paulo one of the world's largest megacities—and its problems are correspondingly large. São Paulo illustrates well the range of Brazil's urban environmental problems, from those that plague the most industrialized countries to those more typical of poorer ones. For the poor, Brazil's urban quality of life arguably lags behind that of countries at similar levels of development (Caccia Bava 2002, 361).

By the beginning of the 1970s metropolitan São Paulo was already one of the most polluted cities in the world, and the São Paulo Secretariat of Health received hundreds of complaints annually about industrial emissions. A master plan in 1968 set out some initial strategies for solving urban problems but could not get traction on its big elements, which included a subway system (Bolaffi 1992) and stimulating growth on an east-west axis (Ferreira 1993, 144). This failed effort coincided in time with the famously successful efforts at urban planning in Curitiba, the largest city in Paraná, just south of São Paulo (see e.g. Menezes 1996; Rabinovitch 1992; Rabinovitch and Leitman 1996; Roberts and Thanos 2003, 112–16). It is worth noting that the population of the São Paulo metropolitan area was already over eight million in 1970, while Curitiba had just 600,000 inhabitants. São Paulo's disorderly development and late planning efforts make the city quite a bit more typical of Brazilian urban politics than Curitiba is.

Ironically, the concern with air pollution in São Paulo was growing at precisely the same time as the Brazilian government was taking a militantly pro-development position at the United Nations Conference on the Human Environment in Stockholm (June 1972). However, the belief displayed by the Brazilian delegation that new technologies could resolve the problems created by older ones was entirely consistent with an acknowledgment that pollution in São Paulo had got out of hand. Institutionally, the solution was to develop a measurement and regulatory capacity together with incentives for firms to install anti-pollution equipment, and sanctions for those that did not.

São Paulo's serious environmental problems are joined with its unquestionable status as Brazil's environmental policy innovator, much like California in the United States (Lowry 1992).² A World Bank study called the state's Environmental Sanitation Technology Company (CETESB) "the premier environmental protection agency in Latin America" and noted that much of Brazil's national-level pollution control legislation originated in São Paulo (Redwood 1993, 70). The Environmental Sanitation Agency (SUSAM), created in April 1970 and located within the Secretariat of Health, obtained sophisticated measuring equipment in 1972 and set up thirty-seven sampling stations in greater São Paulo, twenty-one in greater ABC (the industrial area to the city's south),³ three in Cubatão, and one each in a number of other large cities of the state. Based on these measures, SUSAM closed three industries in São Paulo in 1972—two pesticide plants and a metallurgical one. In February 1973 the São Paulo state government signed an agreement with the United Nations Development Program and the World Health Organization to improve research and pollution control programs in São Paulo. The focus here was on planning—in the words of the superintendent of SUSAM, Nelson Nefussi, "environment should be one of the variables included in planning and in national development programs, so we don't fall into the same mistakes as the developed countries did" (*Folha de São Paulo*, 30 July 1973).

Still, the situation was so bad in 1973 that when the painter Miguel Abellá began to walk around with a gas mask in a silent protest against the lack of attention to air quality, he captured the imagination of many Paulistas. His act led to the formation of one of the first "new environmentalist" groups in São Paulo, called Art and Ecological Thought (MAPE), which led citizens' initiatives in this area for many years (Antuniassi 1989). São Paulo has always had one of the largest concentrations of environmental organizations in Brazil, with 456 in 2002 (IBGE 2004, table 8).

Together these state and nonstate actors have created a series of networks to tackle São Paulo's urban environmental problems. In the 1980s their networks tended to follow the pattern of blocking networks aimed at industrial sources of pollution. Yet even in the middle of a major mobilization against industrial pollution in Cubatão, São Paulo, they began to lay a framework for the largely enabling networks of the 1990s. These helped to develop innovative strategies for dealing with a more complex set of urban environmental problems, strategies that put the urban population at the center as both victim and perpetrator of environmental degradation. We look at two of these strategies: participatory councils for environmental impact assessment and licensing; and automobile policies, both technical and social. We conclude with a section on an emerging Environmental Justice Network, which is tackling some of the questions still unresolved.

There are several themes that run through all of the campaigns in this chapter. Once again, "the state" is not a unified actor but is divided in important ways. Pollution highlights conflicting competencies in a federal system. Since 1981 the regional or state level has been mainly responsible for pollution control, and this has also been true in practice in São Paulo, which has strong traditions of state environmental leadership. However, governing institutions at federal, municipal, and trans-jurisdictional levels (such as the metropolitan level, or that of the river basin) also compete for control on particular matters. Environmental jurisdictions are overlain by different federal arrangements for closely related domains such as transportation and industrial development. All these juxtapositions typically place different state actors on various sides of an issue. Further, there is a regular tension that follows the lines of the debates over technocracy versus democracy, with scientists and technical staff in government agencies often differing from participation-oriented activists in their assessments of causes, solutions, and outcomes of environmental issues. These differences even cut through the middle of an agency like CETESB and divide civil society actors.

Finally, questions about development appear against two backdrops. One is the central debate about what constitutes progress in a country like Brazil: smokestacks and cars, or filters and public transportation? The second is Brazilian development's ugly twin of social and economic inequality, which intersects with environmental questions in the realm that has come to be known as environmental justice (Acselrad, Herculano, and Pádua eds. 2003b). While the label is recent, the phenomenon is not, and points to the greatly increased burden of many pollution problems for poorer urban inhabitants.

Cubatão

The city of Cubatão gained fame in the 1970s as the "Valley of Death," the most polluted city in the world. At the same time, from the standpoint of purely economic and industrialization indicators, Cubatão was a success story on a grand scale. This juxtaposition, and the timing of major revelations about environmental damage there just at the turning point of Brazil's democratization process, make the story of Cubatão a foundational one in Brazilian urban environmental politics. Because of it, the optimistic developmentalist narrative of technological progress came to coexist with a cautionary story about ecological risk. In addition, technocratic approaches to pollution control came into a creative friction with a new politicizing of the environment. In the end, four groups of actors worked together to change Cubatão: local grassroots leaders; their elite environmental allies, including the premier Brazilian scientific organization; a state government won by the opposition MDB party in 1982, the first gubernatorial elections in seventeen years; and staff of the state environmental agency CETESB. The media and World Bank played supporting roles.

Background: Developmentalist Cubatão

Cubatão is located at the base of the coastal palisade known as the Serra do Mar, part of the lowlands around the estuary of the port city of Santos. An area of heavy precipitation and marshy soil, it was not immediately recognized for its assets as a pole of industrial development, except for its location between the city of São Paulo and the Atlantic Ocean. This region of banana growers could never have become an international symbol of industrial pollution, but for the decision in the 1920s by LIGHT (Companhia Light and Power) to make use of the steep declivity from the Planalto to the sea to generate hydroelectric power. The end of the Second World War ushered in an era of nationalist developmentalism grounded in that power. Spurred by the war's lessons in the perils of depending on others for industrial goods, successive governments strove to make the state a motor of growth and national autonomy. In 1951 President Getúlio Vargas pursued rapid industrialization in his second presidency, setting up a national development bank in 1952, and in the following year nationalizing oil production. The formation of Petrobrás, the state oil company, was accompanied by a massive popular campaign around the slogan *o petroleo é nosso* (the oil is ours). The prior administration of President Dutra had already decided to locate the Presidente Bernardes oil refinery in Cubatão. There was abundant energy, and proximity both to the nation's largest energy market and to excellent port facilities made the site appear perfect. At no time, apparently, were either ecological or social characteristics of the area taken into account (Ferreira 1993, 52–53). The refinery began operation in 1955.

Meanwhile, in 1951 a group of private investors led by Plínio de Queirós, a brilliant, civic minded engineer "interested in really contributing to the progress and autonomy of the country" (Branco 1984, 53–54), drew up plans for installing a steel company—Companhia Siderúrgica Paulista, or COSIPA—on almost five square kilometers of landfill and swamp, effectively reinventing the landscape, at considerable cost. By 1961 the original investors had run into financial difficulties, and the National Economic Development Bank (BNDES) assumed a 58 percent share of the project.

With the construction of COSIPA, Cubatão was gradually transformed into a steel and petrochemical development center. Other industries set up shop to use the petroleum derivatives from the refinery. By 1980 Cubatão housed twenty-three major industries and more than eighty small and medium-sized ones,[4] public and private, domestic and international. The population had reached 78,327 (Guilherme 1982, 17). In 1984 the region's industries were responsible for 3 percent of Brazilian GDP and produced 59 percent of its fertilizers, 50 percent of its nitrogen, 47 percent of its styrene, 30 percent of its chlorine, 15 percent of its steel, 14 percent of its gasoline, and 12 percent of its liquefied petroleum gas (*Exame*, 7 August 1985, 37). Collectively these industries launched around a thousand tons of pollutants a day into the air, of which more than 250 tons were suspended particles.

In 1964, after the military coup, Cubatão was declared a strategic area, which brought it under tight federal control. Over the next several years, the political implications of this military classification became clear: the town's mayor would be appointed by the (appointed) governor with the approval of the (military) president; the municipal council would have its prerogatives severely truncated. But the area continued to grow, and jobs were plentiful. By the middle of the 1960s Cubatão's big dreams included nightmares recounted in hushed voices. Everyone had been willing to put up with some hardship in Cubatão, in exchange for the promise of a more secure future that growth and good jobs seemed to offer. But something abnormal was afoot. Rumors flew about the birth of deformed babies, about sick and dying children, contradicting the premise that present sacrifices would earn a better future; instead, they implied that the future was the very thing being sacrificed. Social conditions in the city were appalling. Although it had the highest per capita income of any city in Brazil, 35 percent of the population lived in shantytowns with no social services (*New York Times*, 23 September 1980). Shantytowns were perched in precarious positions on the mountainside, often inside the state park of the Serra do Mar, or were built on stilts in the marshy areas near factories. Only 25 percent of the city's residences had running water, and there was no sewer system (Queiroz Neto, Monteleone Neto, and Marques 1984, 81; see also Lemos 1998).

The "Valley of Death"

Housing conditions were one cause of the startling health situation that brought Cubatão to public attention in the late 1970s. There had already been other indications of trouble. In 1975 researchers at the Oceanographic Institute of the University of São Paulo had detected mercury levels of 2.5 micrograms per liter—twenty-five times the recommended maximum—in the water and sediment of the Santos estuary. By the end of 1977 the damage to surrounding vegetation from factory emissions was becoming visible. In 1978 Florivaldo de Oliveira Cajé, a municipal councilman from the opposition party MDB, claimed that at least one death and many cases of skin lesions had occurred among workers at the Rhodia factory making phenyl pentachloride. In May 1979 Alberto Pessoa de Souza, a municipal councilman from the PDS, began a study of what appeared to be an unusual number of congenital deformities among babies born in Cubatão (Kucinski 1982, 11). His research interested an old school friend Reinaldo Azoubel, whose findings of unusually high rates of anencephalia (babies born without brains) attracted widespread attention and contributed to a growing stigmatization of Cubatão as a place of unnatural occurrences, a place where monsters were born (Ferreira 1993, 113–14).

The international press began to cover Cubatão in some depth around 1980, with articles in 1980–81 in major papers like the *New York Times* and the *Washington Post* and wire services like Reuters and United Press International, which picked up the moniker "Valley of Death" from the local media. In 1981 the state assembly set up a commission of inquiry into Cubatão; the commission reported back that the situation required serious pollution abatement measures, but nothing concrete came of it (Lemos 1998, 80). Instead two opposing movements emerged, each trying to shape interpretations of the developments and responses to them.

For their part, local grassroots actors moved to deepen the critique. Their organization was stimulated first by local opposition politicians, working with local churches to gather information and take up the cases of families whose children had been born with deformities. The Association of Victims of Pollution and Bad Living Conditions of Cubatão was created out of groups in Vila Socó and Vila Parisi, two of the worst slum neighborhoods in the region, directly abutting the factories. Since the victims' association was not allowed to formally register at first, still a necessity under the military government, it is difficult to date its emergence exactly. In the early 1980s the association held neighborhood meetings to gather information on which issues were considered most important to residents. Through marches and manifestos they drew attention to their plight, seeing mass mobilization as their best strategy (Ferreira 1993, 122–27). As Lemos and Looye note (2003, 353–54), the victims' association was critical for reframing pollution as a social problem that required new kinds of development. Nonetheless, its positions were never consensual in Cubatão, where many union members and residents worried that too much talk of pollution might lead to plant closings and to loss of the hardship pay that workers earned for being subjected daily to unhealthy conditions.

Faced with the victims' challenge, the secretary of state for industry and commerce set up the Vale da Vida (Valley of Life) working group at the end of 1981, which included representatives from the industries in the area and was named as part of an effort to negate the rapid dissemination of the name "Valley of Death." Vale da Vida presented its first proposals for solving the pollution problem in December 1981, asserting that all would celebrate the end of pollution in Cubatão by 31 December 1983. The group recommended that residents in the most polluted areas be moved, instead of getting to the source of the problem. Lúcia da Costa Ferreira (1993) associates this "solution" with a desire on the part of local politicians and policy makers to end the widespread association being created in the media between Vila Parisi, the most heavily polluted slum area, and Cubatão itself.

In March 1982 the federal government (still led by the military) set up an interministerial commission with representatives of municipal, state, and federal governments, business associations, and scientific groups. At the same time, at the suggestion of the Vale da Vida group, the state government issued a land use decree, dividing the area into three zones: an industrial zone, a zone to include only nonpolluting industries, and a residential zone. This kind of political and economic coalition had been the hard core behind the developmentalist projects of the last decades. The real novelty was that environmentalists and victims of the model were able to put together a strong counter-network.

The victims' association could depend on elite allies to support its struggle. Randau Marques's articles on Cubatão in the *Jornal da Tarde* in 1979 had amounted to a crusade that had drawn attention to the cause. OIKOS, one of the first professionalized environmental organizations in São Paulo, made pollution in Cubatão the focus of its annual campaign in 1984. Its co-founder Fábio Feldmann was the attorney for the victims' association, helping to obtain legal status when the city government refused to register it. He also brought Cubatão to the attention of the SBPC (Brazilian Society for the Advancement of Science), which created a study group on Cubatão in January 1982. The SBPC proved to be an especially important ally.

The SBPC attended a congress organized in April 1982 by the victims' association, putting its professional standing behind the association's demands. Besides calling for pollution control, the victims' association wanted to see a closer connection between fault and reparations, demanding that fines assessed on industries be used to benefit the communities, and that the industries pay for damage to health. The organization also protested the zoning decree (Decree 15.825 / 82) that called for removing precariously situated neighborhoods, and demanded the right to participate in formulating solutions (Queiroz Neto, Monteleone Neto, and Marques 1984, 82–83; Ferreira 1993, 122–27). Clodowaldo Pavan, a geneticist affiliated with the SBPC, addressed the assembly, celebrating the importance of its actions and radically restructuring the causal story about Cubatão:

> The solution to the environmental problem in Cubatão will also be a solution for Brazil—your success is a success for the Brazilian people. This Association being formed here today is important . . . for the whole country . . . the instrument of a historical period where people began fighting for a more humane way to live. . . .

Anywhere in the world, the environmental situation in Cubatão would be considered a national disaster. This critical situation concerns the world in the same way as Seveso, Bilbao, and Minamata did, and the difference between those events, all of which happened in developed countries, and Cubatão, is precisely the way the problem gets treated. That is what distinguishes real progress, which raises the standard of living of the people who produce it, from that degenerate, short sighted progress that doesn't care about anything but pure and simple profit.

Some people try to justify pollution by saying it's the price we have to pay for progress. These people say, OK, if automobiles kill people, this isn't a reason to take cars off the road. The answer is that a car accident is exactly that—an accident. What we have here in Cubatão . . . is something else. Using the same example, we could say that what is being done in Cubatão would be the equivalent of producing automobiles without brakes on purpose. That is not an accident. (cited in Ferreira 1993, 128–29)

At the 34th Annual Meeting of the SBPC, the Cubatão Study Group organized a symposium to assess "the results of ten years of disastrous environmental policy," which followed the UN conference on the environment held in Stockholm in 1972. The symposium, called "Stockholm 72 x Cubatão 82," placed Brazil's primary academic organization squarely on the side of the population of Cubatão (and in direct confrontation with the federal government). The SBPC played an important role in publicizing the story of Cubatão, especially in relation to the nonspecialists in the victims' association. In a discussion in which the portrayal of issues as highly technical had made it easier to exclude dissenting voices, the scientists lent the unquestioned legitimacy of their credentials to the association. The local population was accustomed to powerlessness, to being treated as incompetent; the SBPC scientists ended their isolation and helped to demystify the terms of the debate (Ferreira 1993, 127–30).

In the process the SBPC became more openly critical of both the military regime and an economic model that fueled growth by squeezing workers and the poor.[5] This more critical posture accompanied a shift in constituency, as the organization opened its ranks to social scientists (although anthropologists had been participating for several years) (Fernandes 1990). The sessions on Cubatão at the 34th Annual Meeting were among the SBPC's most outspoken acts. In response to the study group's report, the SBPC General Assembly passed its first resolution about Cubatão, essentially duplicating the demands of the victims' association. The report further pointed out that none of the official solutions yet proposed addressed the social needs (housing and health) of the population, a situation that seemed particularly unjust when local tax revenues were so high.

Assigning institutional responsibility for finding a solution was especially difficult in Cubatão because of its special status. The city was under federal jurisdiction both as a national security area and as a center of petro- and chloro-chemical production. State and federal governments both had oversight powers on land use, normally a municipal responsibility. Overlapping jurisdictions made it hard to pin blame on anyone. For a long time, official organs simply denied that health statistics were anomalous, arguing that malnutrition was as likely to cause birth defects and infant mortality as pollution was.

Political Opportunities

Brazil was intensely politicized at the end of the 1970s. Hoping to stem a recent tide of opposition party victories in elections, in 1979 the military government changed the rules of the electoral game. Subsequently, party and electoral politics became much more important, as already discussed in earlier chapters. The elections of 1982 marked the first time governors had been elected directly since 1965. Also being elected were municipal, state, and national legislative bodies and most mayors. The campaigns intensified the level of attention to the Cubatão case during 1982. However, they also helped demobilize the alliance that had initially fortified the victims' association. Dojival Vieira dos Santos, principal organizer of the association, was elected to the municipal council for the PT. The local Catholic Church, which had supported the movement, was run by a priest with strong PMDB sympathies, who therefore split with Dojival, as did the leadership of the SBPC, which supported the PMDB candidate Franco Montoro for governor.[6] Besides the partisan rifts, demobilization also came from the expectation that with former opposition figures now in government, those seeking change would have new opportunities for dialogue and access.

Montoro's election to the governorship of São Paulo in November 1982 eventually did precipitate a change of course on Cubatão. For over a year, the new administration surveyed pollution sources and studied potential solutions. Montoro made the Cubatão cleanup a priority, a decision that its international visibility and local opposition made hard to avoid. By appointing Werner Zulauf president of CETESB, the state pollution agency, he was ensuring that his government had a negotiator whom the main players in Cubatão could accept. Zulauf had been an earlier director of CETESB, was a member of the Vale da Vida commission, and had worked for COSIPA. He was acceptable to both the local government in Cubatão and the industrialists in the area.

CETESB had had "winter operation" technical programs in place since 1976 that provided for the use of lower-sulfur fuels and possible production shutdowns during times of critical air conditions from May to August (SMA 1997b). Under the new government CETESB added a program of community participation and education to its oversight of pollution sources and its incentives to reduce pollution (CETESB 1984, 1). The new programs were capable of being implemented with local populations even as their leadership dissolved. This mix of technical and community-based programs was an innovation for CETESB, for Brazil, and unusual even worldwide at that point. Zulauf reflected on some of the difficulties of that transition in an interview in 1991: "There was a lot of participation in this process but also less than people think. People thought the population of Cubatão was more concerned with the problem and basically they were worried about jobs. The press stirred up community mobilization a bit because there was a lot of misinformation. There was a lot of activity by the leadership. Ecological leadership has this role—to mobilize the community, raise its consciousness, raise problems. The role of technical people is to solve them. If a community leader uses the wrong language or concepts this isn't important for the listener, who doesn't know much about these things either. This is strange for the technical person. We need to understand the different roles of each, because if it weren't for the environmentalist leaders there would be no political action to defend the environment in the world today" (interview with Zulauf 1991). Many of the meetings reflected a sense that residents had a right to be informed of technical measures, but without giving them much active participation except for that afforded by a complaint line. Nonetheless, CETESB implemented its programs enthusiastically enough that industrial actors chafed under the new controls, with their representatives comparing CETESB to a traffic cop—and one that had taken sides against them, with the community (CETESB 1984, 5–6).

In early 1984 industrialists decisively lost their ability to define the terms of debate on Cubatão. On 25 February 1984 a leak in a Petrobrás pipeline (reported to the company the day before) ruptured, with an explosion "like a burst of gunpowder" (*Washington Post*, 26 February 1984). At least eighty-three people in the Vila Socó shantytown died, and some estimates ran well into the hundreds, as the fire burned hot enough to incinerate bodies (*New York Times*, 26 March 1984). Over the next year pollution emergencies, forced evacuations due to chemical leaks, and the added trauma of the disaster in Bhopal, India, in December 1984 (caused by leakage of a chemical that Union Carbide was also using in its plant in Cubatão) decisively shifted the focus from removing the stigma to removing the risk.

At this stage, with the state government fully committed to the policy, CETESB began to negotiate technical solutions and deadlines with the industries. The negotiations followed the industries' request for positive assistance in overcoming their pollution problems rather than fines (CETESB 1984), one of Brazil's first policy initiatives in this direction. The Montoro government renegotiated the terms of a World Bank loan for $58 million (Loan 1822-BR) initially intended to close out in March 1984. The loan was approved initially on 27 March 1980 to help finance the São Paulo Industrial Pollution Control Project and came into effect in January 1981. Intended to reduce particulate emissions, treat liquid effluents, and provide technical assistance to CETESB, it had run into problems regarding counterpart funding and an initial lack of demand by eligible industries. In September 1984 the loan was reformulated and reduced to $34 million; its new closing date was set at 30 June 1986. The bank's share of subprojects went from 30 percent to 50 percent, emissions were expanded to include sulfur dioxide and toxic waste, and more flexibility was built into the procedures for approving projects and loan subconditions. Most especially, the geographic scope of the loan was expanded to include the whole state, with emphasis on Cubatão (World Bank 1991). For a World Bank analyst, the reformulated Cubatão project was the most successful of the bank's early environmental projects in Brazil (Redwood 1993). The bank itself notes that its role was secondary: "the critical role of local public awareness and unequivocal political support for CETESB's initiatives cannot be overemphasized" (Redwood 1993, 47). In fact the World Bank's credit line was only "catalytic," with most companies motivated sufficiently by fines and negative publicity to install pollution controls using other resources (Redwood 1993, 47).

Zulauf concurred that publicity was one of his main assets in carrying out policy in Cubatão, and in creating the political will in the government to make it work. He saw his main allies as the press (international, national, and local), public opinion, and environmental organizations: "Political will can exist spontaneously or be induced . . . Today, most of our politicians have no commitment to the environment, except sometimes via discourse, or opportunistically. If we had to wait for the political will of governors to arise naturally, we'd be in bad shape. The other way is to induce it by showing they'll gain authority by forming this alliance with public opinion, the press, environmental organizations, state technical personnel, and some economic actors involved in pollution control" (interview with Zulauf 1991). This new understanding of the important role of the public laid the groundwork for building enabling networks and fit well with the larger process of transition to civilian rule.

Assessing the Legacy of Cubatão

Environmental impacts can be assessed in at least two ways. To measure political effectiveness, one asks whether environmental policies have been agreed and put in place; to measure environmental effectiveness, one looks for impact on physical indicators of environmental quality (Kütting 1999). By both measures the Cubatão legacy is a comparatively positive one, although both also indicate certain gaps in effectiveness that needed to be resolved with a new generation of urban policies in the 1990s.

The pollution crisis in Cubatão produced an unusual degree of political consensus in favor of environmental protection in São Paulo in the 1980s. State political leaders, the state environmental bureaucracy, environmental and scientific groups, and grassroots community members participated in an unprecedented joint effort able to overcome developmentalist patterns that had lasted half a century. They were helped by graphic problems such as birth defects and destructive fires that won broad support for their cause and enabled them to link social and environmental concerns in a common critique of the existing development model. Together they wrote more stringent regulations and held polluters to account, offering them both positive incentives and the threat of punishment. The regulatory model continues, as do a few pieces of community oversight, such as a complaint hotline. CETESB continues to seek the "unrestricted vigilance" of the Cubatão community and its elected officials to keep CETESB apprised of pollution problems, and the population has complied by making numerous calls, 376 in 1993 (CETESB 1994, 3). Specific institutional arrangements made during these years for community education and involvement have largely disappeared, lacking the conditions to sustain themselves (Lemos and Looye 2003). However, as we show below, other state-society linkages were forged in the 1990s, performing similar enabling functions.

Environmentalists themselves have largely moved on to other issues. Of 439 organizations registered with the State Secretariat of the Environment (Secretaria do Meio Ambiente; SMA) in São Paulo in 2004, only five indicated that they regularly worked on some aspect of industrial pollution and three more worked on air quality.[7] The general move away from protest activities is evident in the prevalence of environmental education, recycling, and tree-planting projects among the registered groups. Despite these developments, a broad array of general-purpose organizations and environmentalists in university and government positions is poised for action on pollution issues, and a new Environmental Justice Network, discussed at the end of this chapter, includes some of them. Many participants in this network are not registered with the secretariat, since they do not think of themselves as environmental organizations but as unions or other social groups.

Conceived to limit fixed-point industrial pollution in Cubatão, the policies achieved mixed levels of environmental effectiveness. Positive evidence includes a reduction in really severe levels of pollution. In 1984 CETESB began to implement the Plan for Prevention of Extreme Episodes of Air Pollution. States of alert in Vila Parisi continued in the plan's early years (one state of alert in 1986, four in 1987, and three in 1988), but there was only one between 1992 and 1997, in 1994 (CETESB 2000, c). The only state of emergency (the highest level) in the 1990s resulted from that same episode, but CETESB concluded that it had meteorological rather than emissions-based causes (Romano, Alonso, Lacava, and Martins 1995).[8] In 1999 the German bank KFM joined with the state environmental secretariat to implement a $440 million plan for restoring the health of the Serra do Mar State Park near Cubatão. The plan provided for the state construction of low-income housing (*A Tribuna*, 11 July 1999). COSIPA, "an old villain from the black days of Cubatão," had remade its image by 2000, investing $200 million in improving the air, water, and soil at its company location and signing up for voluntary ISO 14001 environmental certification (*Veja*, 17 May 2000, 177). Oversight of the São Paulo industries continues in a routinized way, with standard procedures for attacking potential emissions problems.

Against this picture of success, others found evidence of failure, especially in the mid-1990s. In 1994 CETESB acknowledged that its equipment was old and failing (*Estado de São Paulo*, 4 September 1994), and its air quality data developed gaps until the equipment was replaced under the next administration. Even then Greenpeace and the Group for the Study of Air Pollution at the University of São Paulo used their own equipment in 1996 to measure emissions and concluded that the fixed monitoring stations used by CETESB failed to monitor critical new air pollution areas. For the NGOs the bigger problem was that the information CETESB was providing was not enough, or not good enough, for the population to judge the seriousness of air pollution in their city (Jacobi 1997, 19–21). Greenpeace and others also tried to persuade CETESB to make the emissions standards more stringent, but CETESB defended them as meeting World Health Organization guidelines (Jacobi ed. 1997, 23); at the national level CONAMA also declined to review the issue, arguing that it was a state-level matter (interview with Rodrigues 2000). More recently a larger group of environmentalists charged that although companies must put pollution controls in place under CETESB's licensing process, the agency's follow-up monitoring has been too weak to verify implementation (Svirsky ed. 2002). CETESB publicly admitted that this was a flaw in its oversight (*Folha de São Paulo*, 5 October 1998) which has not been corrected in subsequent years of budget austerity. Finally, as existing industrial sources of pollution were controlled, air quality in São Paulo increasingly depended on two kinds of more cooperative, long-term efforts. First, there needed to be procedures making it possible to foresee and possibly abate industrial pollution before it reached the emergency levels of Cubatão. (Below we discuss how São Paulo has met the challenge of environmental licensing.) At the same time, since the mid-1980s, and partially as a result of these policies, industrial sources have played a steadily smaller role in regional air quality. Heavy smog still hangs over the São Paulo metropolitan area—smog that is even more evident as one sits in the interminable traffic snarls that make driving there so stressful. Emissions from personal automobiles and related pollutants such as those from tires and fueling have come to play by far the largest role in air pollution in metropolitan São Paulo (CETESB 2000, 4). These diffuse emissions present much more complicated challenges for policy makers, and CETESB has struggled to find solutions to a problem whose complex causes fall largely outside its control.

Institutionalizing Pollution Control through Councils

In 1984, just as the Cubatão crisis peaked and attention began to turn elsewhere, a series of administrative changes began to rework the politics of pollution control in Brazil at both the national and state levels. Unlike the fires and birth defects that grabbed international and domestic attention for Cubatão, these changes went largely unremarked except among the political communities directly linked to environmental issues—and the business firms they monitored. Conceptually, the changes marked the transition from a blocking to an enabling strategy on the part of environmental actors, albeit building on some of the alliances initially constructed to stop pollution in Cubatão. Chapter 2 introduced the CONAMA, the National Council on the Environment, which set up the framework for this transition. The state of São Paulo created its own CONSEMA, or State Council on the Environment, in 1983. Together these primarily governmental actors shaped a new preventive approach to pollution control that continues to this day.

The national legislation that created CONAMA in 1981 foresaw the requirement of environmental impact assessments, and CONAMA wrote a directive imposing the requirement in 1986. While the national body established the general requirement and procedures, the assessments themselves took place at the site of the potentially polluting activities, most of them initially under state jurisdiction and, after 1998, also at the municipal level. As we have stressed repeatedly, a mandate for subnational action will have very different results in different parts of Brazil, and São Paulo and more urbanized locations have generally had more effective licensing processes (Glasson and Salvador 2000). The comparative success in São Paulo can be traced to the activities of its state-level council, the CONSEMA. While imperfect, CONSEMA's evaluations of environmental impact assessments and reports have partially transformed the politics of pollution into a routine affair, with work carried out in advance to curb at least some of the problems that Cubatão exemplified.

A variety of nonstate actors have been an intimate part of that process, marking it as a significant enabling activity that crosses state-society lines. Born in 1983 out of the democratizing wave that brought Montoro's opposition gubernatorial administration to power, CONSEMA was set up with five (now six) representatives selected by environmentalists, and an array of representatives of researchers, universities, and the legal community. The eighteen nonstate representatives have formal parity with the eighteen governmental representatives (Svirsky ed. 2002, 53). In its early years CONSEMA interpreted its mandate in sweeping terms. It felt free to send letters of reprimand to the national government for perceived environmental infractions, and even chided the Soviet government after the nuclear disaster of Chernobyl (CONSEMA 1993). No one within the state of São Paulo escaped its scrutiny either. Perhaps most importantly, CONSEMA began to carefully examine proposals for environmental licensing as well as proposed environmental policies, with CETESB's technical skills and experience anchoring the process. CETESB had begun environmental licensing in 1976 (CETESB 1998c), but the process was expanded and opened up with CONSEMA. After Montoro left office CONSEMA was trimmed back to primarily this technical function, and it continues in that form now. The transition was a hard one for environmentalists, who wanted CONSEMA to play a larger political role in exchange for the time and effort of working with it.[9] At the same time, representatives of technical agencies found many of the technical discussions marred by extremely politicized participation (Furriela 1997). CONSEMA has continued to play an important role from the standpoint of pollution control. In a backhanded compliment to the effectiveness of CONSEMA in this regard, the developmentalist mayor of São Paulo, Paulo Maluf, created a municipal-level council to create a quasi-legal alternative licensing channel for projects that he knew CONSEMA would block (Svirsky ed. 2002, 102).[10]

São Paulo state has recently moved ahead of the national pattern to revitalize and modernize its licensing process. This time the changes have been justified as part of a unilateral move on climate change issues, with São Paulo joining California in an effort to lead their recalcitrant national governments into action (Reid, Lucon, Coelho, and Guardabassi 2005, 18). In 2002 a state decree made environmental licenses for stationary pollution sources good for only five years, thus prodding long-licensed companies to update their emission controls. The government will monitor total emissions in different areas, and it will issue abatement credits for any reductions. In areas of poor air quality there will be no new licenses unless sufficient abatement credits are available. Trading of the credits will be done independently in the private sector. These licensing policies are joined with landfill and transportation emissions plans, among others, in São Paulo's State Program for Climate Change (Reid, Lucon, Coelho, and Guardabassi 2005, 16–19). The transportation initiatives turn our attention to what has become São Paulo's largest source of air pollution: private vehicles, and especially the private automobile.

The Automobile, Pollution Policy, and Social Inequality

For the individual, the automobile symbolizes many of the same things that industrial development does for a country. Both embody modern ideas of progress and are tangible symbols of wealth. Both promise to speed up the basic tasks of daily life and so to make many more options available, and the automobile literally carries the individual to those possibilities. The automobile, unlike public transportation, takes the individual to whatever location desired, at whatever time. The individual can choose who accompanies him or her on this journey, pick the route, select the soundtrack. In the words of Feldmann, cars represent "power, prestige and liberty" (Introduction to SMA 1997a). This vision of the automobile as the embodiment of individual freedom and progress, as powerful in Brazil as it is in the United States, has to be placed in a larger context, however.

The environment in which people drive their cars is circumscribed, organized and disorganized, by public and private choices that precede a motorist's decision to pop over to the grocery store. Where the roads are laid down, how they are maintained, what destinations are available, and whether there will be parking are just a few of those. Prior choices also determine whether other forms of transportation might be used instead, and if so what their cost and quality are. These spatial parameters confront not just the one individual but all of a city's inhabitants and visitors, who are making simultaneous choices about how to move about. An individual impulse to go to the beach on a nice day is likely to result in a collective traffic jam. The 5.5 million cars of metropolitan São Paulo (CETESB 2000, 3) overwhelm individual freedom to move, with peak time traffic crawling at twenty kilometers an hour in 1994, down from twenty-eight kilometers an hour a decade before (Rodrigues 1996, 6) and certainly even slower a decade later. The rise in the number of cars has occurred very quickly, as there were only about two million cars in 1985 (CETESB 1985). The importance of these considerations is obvious when one compares the situation in São Paulo with the much more orderly one in Curitiba, where the public transit system is so efficient at moving people from home to workplace that three-quarters of the population uses it, including many car owners (Rabinovitch and Leitman 1996).

On the other hand, not all people have the same choices, and the line between those who do and those who do not have cars is increasingly important in cities like São Paulo for reinforcing socio-economic divisions of various kinds. The automobile is a moving "fortified enclave" (the term is from Caldeira 2000, 4) that allows those who can afford one to move more freely, safely, and—still—more quickly around the city than their fellow residents on foot or using public transportation. Private cars allow the middle and upper classes to avoid setting foot on streets and other public spaces as they move from their comparatively secure homes to their comparatively secure jobs, shops, and leisure activities, all in spaces that are increasingly private as well.

The class dimension of the private automobile in São Paulo is made clear by looking at the distribution of vehicles and housing types. Since the 1940s São Paulo has had its middle and upper classes in the central part of the city and the poor and working classes relegated to the periphery (Caldeira 2000, 220ff). This fact lends special significance to the distribution of cars: the central municipality of São Paulo has about five million vehicles for a population of about 9.5–10 million people, while the additional 6–7 million people in the surrounding metropolitan region add only another one million vehicles (interview with Steinbaum 1999). Despite being farther from the jobs, shops, and government offices in the center city, it is the peripheral populations who depend on a precarious public transportation system and even walking to reach those central locations. Even Curitiba shares these problems, with troubled unplanned settlements ringing its exemplary core (Macedo 2004, 546–48).

The peripheral locations with limited public services are also disproportionately the homes of darker-skinned Brazilians (Paixão 2003). In sum, the existence of vehicle-based pollution—currently the major source of all air pollution in metropolitan São Paulo—is a classic issue of environmental justice: a relatively well-off few make private choices that externalize the costs of their choices to wider society—with the twist that these few cannot escape the negative costs in health and environment either. This reality is just starting to be acknowledged in the Brazilian environmental community (see Acselrad, Herculano, and Pádua eds. 2003a), but it obviously has roots in the Cubatão case as well.

Most of the policies followed by CETESB and SMA to reduce air pollution from vehicles are quite unrelated to the social dimension of the automobile. Networks of state officials have developed a series of policies focused on the physical automobile itself that use both regulation and incentives to reduce emissions, in the belief that these policies are the most effective for confronting vehicle-based air pollution. Alternative proposals based on much broader social concerns and networks emerged primarily during Feldmann's stint as secretary of the environment and from the environmental community after he left office in 1998. In part because they define the scope of the problem so broadly, the effectiveness of these programs is very hard to measure.

Technical Solutions to Vehicle-Based Air Pollution

Recent vehicular pollution policy in Brazil has focused on two especially serious contaminants that create air pollution problems—tiny particulates from diesel engines such as those in buses and trucks; and carbon-based emissions, largely from passenger automobiles. Broadly speaking, technical solutions to these problems can be found by improving fuels and engines for processing fuels.[11] Evaluating fuels requires considering how they contribute to emissions as they are burned to move vehicles, but also the environmental and social conditions of their production. The engines need to be evaluated both at the point of sale and also as they are used, since poor maintenance and deliberate modification (usually for greater engine power) can outweigh sound design. The appearance of black smoke from tailpipes is a good on-the-street indicator that an engine has been modified or poorly maintained, as are inspections of the engines themselves. Human health concerns have been a continuous rationale over thirty years for combating air pollution (CETESB 1998a), with climate change and global warming an emerging rationale in the 1990s (Diniz 2000; Reid, Lucon, Coelho, and Guardabassi 2005).

Non-Fossil Fuels

Brazil has been a long-time leader in developing non-fossil fuels, with use of the renewable, biofuel ethanol (known in Brazil as alcohol) dating back to the 1920s (Brilhante 1997, 436). Alcohol production gained a major boost in 1975, when the military government introduced an incentive program for producing and then consuming alcohol as a fuel. The primary motivation for the National Alcohol Program (PROALCOOL) was to help Brazil escape from its dependence on foreign oil—with imports then accounting for 83 percent of its total petroleum use and the bill adding $4 billion a year to its mushrooming foreign debt (*Business Week*, 7 November 1977, 84). The threat to national sovereignty from this dependence was part of the military's calculation, as was support for the important sugarcane industry in a time of low global prices (Oliveira, Vaughan, and Rykiel 2005). Thanks to the subsidies and incentives of PROALCOOL, 95 percent of all the vehicles sold in 1984 ran on alcohol, and 43 percent of all the gasoline and alcohol consumed that year was alcohol (Brilhante 1997, 437). Through the 1980s and into the 1990s both the supply and demand for alcohol as a fuel fluctuated wildly, buffeted by shifting global markets for both sugar and oil and by constantly reformulated government policies. At the same time, fuels were regulated to eliminate lead by 1991 and to gradually lower the levels of sulfur in diesel fuel to less than 0.2 percent in the metropolitan area during the winter (SMA 1997b, 132–33; CETESB 1998a).

Finally, in the mid-1990s PROALCOOL was reconsidered one last time, gaining explicit environmental justification. The basic package that continues is one of market-established prices for alcohol, gasoline, and sugar, along with a special investment program to encourage continued development of the alcohol industry (Brilhante 1997, 447–48). One specific aim of the reformulated PROALCOOL was to spur the auto industry to develop vehicles that could also use alcohol and natural gas. The first of these "flexible-fuel" cars—able to use normal gasoline and alcohol interchangeably—was brought into production by Volkswagen in 2003, and other manufacturers then followed. Flexible-fuel vehicles have quickly come to dominate the market, accounting for an estimated 70 percent of new car sales in 2006, and lots of international attention at a time of record oil and gasoline prices (*New York Times*, 10 April 2006).

One recent estimate concludes that in São Paulo state alone, the use of ethanol as a fuel avoided the emission of 82 million tons of carbon dioxide equivalent from 1980 to 2003 (Reid, Lucon, Coelho, and Guardabassi 2005, 14). In this study of ways that California and São Paulo can lead each other to stronger, immediate action on climate change, São Paulo's experience with alternative fuels is to be one of its major contributions to the initiative (Reid, Lucon, Coelho, and Guardabassi 2005, 33). There is no question that using alcohol rather than fossil fuels reduces carbon dioxide emissions, and Brazil's sugar-based ethanol is much more efficient as a fuel than the corn-based ethanol used in the United States. Brazil's alcohol production can even contribute to comparatively clean and renewable energy generation as the wastes are processed. Once the environmental costs of soil erosion and biodiversity loss are taken into account, however, the overall balance tips, and there is a substantial net environmental loss from alcohol production (Oliveira, Vaughan, and Rykiel 2005). There are also significant social costs associated with sugarcane cultivation, which is still largely seasonal manual labor (Brilhante 1997, 445). As with other issues discussed in this chapter, there is thus a strong equity component: many of the benefits of alcohol use go to the residents and especially automobile drivers of the center city, while the environmental and social costs tend to be concentrated at the site of production, in the countryside. This balance is not unlike that of the petroleum industry, itself hardly a model of environmental and social welfare solutions (see e.g. Lieders 2001; Randall 1992).

Emissions Controls

In 1963 Brazil introduced the first standards for emissions from new cars (interview with Steinbaum 1999). The next efforts to monitor vehicular sources of pollution in São Paulo date to 1976, when CETESB technicians began wintertime inspections of the buses operating in the central city to see if their engines were properly maintained (SMA 1997b). More extensive initiatives began in the mid-1980s, when the partial abatement of industrial pollution in Cubatão allowed more attention to these less visible problems. In 1987 CONAMA set up a new program, PROCONVE (Program for Control of Air Pollution from Automobiles; Resolution 18/86, later Federal Law 8723/93) to monitor pollution and noise emissions from new vehicles. This program was largely formulated and then implemented by CETESB, which annually assesses all new models of national and imported vehicles to be sold in Brazil. The standards reached international levels by 1997, with automobile emissions limits on a par with those of the United States and Japan and those for diesel engines tracking three years behind European standards (SMA 1997b, 131–32).

These policies are the primary ones discussed by CETESB technicians responsible for vehicle-based air pollution control, and they dominate the agency's publications. CETESB is convinced of the environmental effectiveness of these programs. The new-vehicle programs in particular reduced 90 percent of polluting emissions in light cars between 1989 and 1997, and 50 percent for heavy vehicles (SMA 1997b, 132). CETESB staff has tracked down many violators of their policies, such as companies operating diesel vehicles that emit black smoke. Fines for black smoke were greatly increased in the mid-1990s, as were inspections—75,000 diesel vehicles were fined in the first eleven months of 1996, more than in the years 1989–95 together (SMA 1997b, 129–31). CETESB also caught the Fiat corporation masking emissions in two of its models, and the SMA saw the case through the courts until a fine of $4 million was imposed—although it needed to invoke the assistance of both CONAMA and the federal Ministério Público to keep the case moving (*Estado de São Paulo*, 4 July 1997).

Several problems limit the effectiveness of these programs, however, and not all have evident technical solutions. One obvious problem with focusing on new vehicle emissions is that it leaves older vehicles unaffected. In 1998 in metropolitan São Paulo 41.9 percent of all cars were made before 1985, meaning that they are untouched by the world-class standards applied to new cars (CETESB 1998a, 5). This part of the fleet produced a disproportionate 73.8 percent of total carbon monoxide emissions (CETESB 1998a, 6). The old vehicle fleet is almost certainly linked to incomes in Brazil, which are also well below world-class standards. Proposals to create tax incentives for replacing older vehicles have received widespread support, but they have foundered on the fiscal crisis of the national and state governments (*Gazeta Mercantil*, 7 May 1999; *Jornal do Brasil*, 27 April 2000).

Another solution would be to begin inspecting vehicles for in-use emissions as well as unlawful changes in engines, such as the dismantling of catalytic converters. CONAMA passed Resolution 7/93 in 1993, which establishes inspections of gas-powered vehicles, and Resolutions 251/99 and 252/99 in 1999 to cover inspections of diesel vehicles. Rio de Janeiro's state environmental agency, which was carrying out inspections of its cars by 1997, found that 40 percent of cars failed the inspections in the first experimental tests (*Globo*, 2 July 1997; interview with Steinbaum 1999). In São Paulo, by contrast, state and municipality were locked in a competition over who would inspect the five million cars in greater São Paulo, delaying the onset of inspections for years. For CETESB the issue was competence in a federal and technical sense—it had the technical ability to oversee inspections of cars, would do so for the other thirty-eight municipalities of the metropolitan region and the rest of the state, and had constitutional responsibility for controlling air pollution. By 1996 CETESB had trained and certified 150 mechanics' offices to do the inspections (Jacobi ed. 1997, 29).

For the municipality the key issue was quite clearly financial: Paulo Maluf's administration wanted to collect the fees for the inspections, and had already awarded the contract in 1995 in a process that a judge eventually canceled because of favoritism (*Diário Grande ABC*, 7 July 1999; see also Jacobi ed. 1997, 30). Nationwide, revenues from vehicle inspections were expected to total almost $1 billion, with about half that amount going to expenses, so this was a prize worth fighting for (*Estado de São Paulo*, 2 August 1999). Through the 1990s São Paulo state and municipality fought this battle through competing CONAMA resolutions, with one in 1999 giving states the power to do inspections everywhere except in municipalities with over three million cars—in other words, in São Paulo alone. The battle is still going on in the courts.

Further complicating matters is the staggering increase in the size of the metropolitan fleet, from about two million cars in 1985 to six million in 1999. This increase has outpaced even the rapid population rise from thirteen to seventeen million people, and miles driven and time spent have risen even more. In 1967, 25.9 percent of all trips in the metropolitan region were made by car, a percentage that rose to 41.9 in 1987 and continues rising (Jacobi ed. 1997, 9). Increases on this scale are not unique to São Paulo but are magnified there. No clear technical solution to this problem exists, and certainly none within the control of São Paulo's environmental agencies. Dealing with the volume of traffic and the reasons why people drive cars returns us to the social dimensions that we discussed earlier. Recognizing that technical solutions alone would not solve the problem, the state environmental secretariat took the initiative to develop a participatory approach.

"Rodízio, an Exercise in Environmental Citizenship"

Fábio Feldmann has already appeared in this book as a lawyer for the victims' association in Cubatão, founder of one of Brazil's first professionalized environmental groups, OIKOS, and the lone representative of the environmental movement to the Constituent Assembly, reelected twice as a representative from São Paulo for the PMDB and PSDB to the National Congress. He served as state secretary of the environment in São Paulo from 1995 to 1998, under the first administration of the PSDB governor Mário Covas. Feldmann was an extremely active secretary, and the record of his term in the secretariat disproportionately fills the CETESB library, especially the bound volumes of newspaper clippings about state-level environmental issues. Both passionate and detail-oriented, inspiring and abrasive, he briefly made the political and social face of environmental issues overshadow the technical one in São Paulo. The policy which best exemplified this period was that of *rodízio* (rotation). Nominally a straightforward volume-control policy already used in Mexico City, Santiago, and other cities to restrict the use of some part of the auto fleet each day, the version deployed in São Paulo attracted interest because of Feldmann's insistence on presenting it as an exercise in environmental citizenship (SMA 1996). Ultimately the purpose of the rodízio was to change the everyday behavior and attitudes of São Paulo's residents, and Feldmann believed this could best be done through education and persuasion.

Attitudes at the start of the rodízio offered mixed signals concerning the difficulty of the task ahead. In a survey of a thousand residents of São Paulo in the early 1990s, Jacobi found that air pollution was perceived as the main neighborhood-level environmental problem by more respondents—13.3 percent—than any other, and that wealthier groups saw it as more serious than poorer groups did (Jacobi 1999, 39). This class difference was also evident when respondents were asked to identify the types of air pollution that affected them most strongly: auto emissions and soot were cited by wealthier groups, dust and bad smells by poorer groups (Jacobi 1999, 58). All groups were most concerned about the impacts of air pollution on their daily lives: dirty houses, irritated eyes, and so on (Jacobi 1999, 62–63). Perhaps most relevant for Feldmann's plan was the broad agreement by all sectors, 89.1 percent of respondents, that government action was best able to solve air pollution problems, with community action cited by only 7 percent and individuals by 3.9 percent (Jacobi 1999, 102). As for specific measures to be taken, 20.9 percent of respondents believed that air pollution should be handled by controlling industrial emissions and 18.7 percent cited vehicular emissions; only 9 percent thought that reducing the number of cars would help (Jacobi 1999, 119).

Against these views, which matched the history of air pollution policy to this point, Feldmann launched his effort to persuade the citizens of São Paulo that they could change their environmental future through the coordinated action of state and society. The first rodízio in winter 1995 was a haphazard affair, launched in part because a journalist picked up on a chance remark by Feldmann during his speech on taking office. This rodízio was a voluntary one, and the environmental agencies were pleasantly surprised to find that 45 percent of the population adhered to the plan—and then surprised again to find that most journalists labeled the outcome disappointing (Medina 1998, 198). Before the next year's campaign, the secretariat went into mobilization mode, calling upon existing networks for support.

For the rodízio to go forward as a set of mandatory restrictions on driving on designated days, Feldmann needed a law, and he needed to mobilize both the state environmental agencies and the population. At the largest scale, Feldmann invoked global problems like climate change and global atmosphere agreements as motivations for addressing air pollution (SMA 1997b). These arguments helped to mobilize environmental organizations, which in turn influenced legislators and SMA and CETESB to act. A petition drive signed by a hundred environmental organizations in favor of the rodízio helped to persuade state congressional representatives to vote for the plan (SMA 1997a, 78). The positive votes for the bill came from parties on the center and right, even though environmental organizations have been historically more closely linked to leftist parties. The bill, sponsored by the PSDB, was largely opposed by the PT (interview with Svirsky 1999). The law passed in 1996 but only authorized a rodízio for two years.

The street activism of Greenpeace-Brazil, which launched a major campaign on air pollution at this time (Rodrigues 1996), helped to convince the environmental agencies that additional action had to be taken (SMA 1997a). Greenpeace's slogan on the rodízio was "Rodízio now! But it's not enough . . ." The organization hit CETESB especially hard during this period for failing to provide adequate public information on the health risks of air pollution, even seeking to sue the head of the air pollution division (interview with Rodrigues 2000). Greenpeace also collaborated with the environmental agencies, joining in a set of weekly discussions at the secretariat in May 1996 that included representatives of federal, state, and municipal government in environmental, transport, and health divisions to discuss sustainable transport policy (SMA 1997a).

The environmental secretariat and CETESB mobilized their own staff to carry out an extensive public environmental education campaign about the rodízio, its rules, and its rationale. They went to schools, handed out pamphlets in the streets and at shopping centers, recruited popular actors, models, singers, and athletes to record public service announcements, and blanketed the media. The numbers are astonishing: from 125,000 calls to informational hotlines in 1995, the number went up to 3.2 million in 1996 (SMA 1997a, 86). In 1996 the air pollution technicians ran a three-day seminar for journalists on air pollution, which resulted in fourteen hours of television coverage of the rodízio in 1996 alone and 120 interviews with Feldmann (SMA 1997a, 106, 107). Even with the rodízio well launched, there were 1,966 articles on it from February to August 1997, with 43.45 percent positive toward the policy, 41.35 percent negative, and only 15.2 percent neutral (Medina 1998, 111). Feldmann criticized the news coverage for presenting overly simplistic views and for decrying the lack of adequate public transportation without noting that it was the responsibility of the municipal government, not the state environmental secretariat, to provide it (Medina 1998, 197ff). From the standpoint of ordinary citizens, however, these issues were clearly related.

Was the rodízio an effective policy? The residents of São Paulo gradually did come to see its usefulness, a relevant measure of political effectiveness; 90 percent of those who did not regularly use cars strongly supported the policy. Regular drivers were less fully convinced: 55 percent thought in July of 1997 that the rodízio was the best way to solve air pollution problems in São Paulo. The numbers for those who believed it did not help significantly dropped from 62 percent in June to 27 percent in July, clearly because of the improvement in traffic flow (*Folha de São Paulo*, 7 July 1997). Similar conclusions emerged from open-ended focus-group discussions carried out for the secretariat by a consulting firm. Drivers started out against the rodízio when they heard the topic, while non-drivers were favorable. After discussion there was a general agreement that "the rodízio is seen as a possibility for real exercise of citizenship: participating with your share of sacrifice and effort to improve the general quality of life, the Citizen forms a partnership with the Public Authority, and feels honored and improved by this" (LPM 1998, 8).

An interesting twist is that citizens also engaged in enforcing the law, a very unusual occurrence in Brazilian public policy. People who took their cars out on forbidden days got "gestures, words, little arguments, expressions of disdain, insistent honking" from their fellows (LPM 1998, 16). The catch was that this attitude depended on a feeling that the public authority was doing its fair share, which participants defined as, for example, making real improvements in public transportation (LPM 1998, 9). For the ordinary consumer, it mattered little that different branches of government were responsible for cars and buses. If commuters were required not to drive on certain days, they wanted alternative forms of transport. Given government failures in this respect, the predominant feeling was irritation: "Life in São Paulo already is difficult and stressful, the traffic is chaotic and drives you crazy, coming and going to carry out normal activities is exhausting and wears you out" (LPM 1998, 12).

Environmentalists agreed completely. This was also the heart of Greenpeace's critique that "the rodízio is not enough . . ." Greenpeace had a long list of legal directives that the municipality was not following, most having to do with improving bus traffic. The organization met almost daily with the Ministério Público, until a change of priorities and reduced funds at Greenpeace International led it to focus on other issues like deforestation in the Amazon instead. The former air pollution campaigner Rodrigues regretted environmentalists' failure to seek commitments from mayoral candidates to work on transportation issues (interview with Rodrigues 2000). Feldmann also recognized retrospectively that the distribution of relevant responsibilities over different political levels and his inability to address central concerns like public transportation made air pollution and other problems very difficult to solve (interview with Feldmann 1999).

In 1998 the rodízio clearly lost its shaky political foundation. The most dramatic outcome was that Feldmann failed to be reelected to the National Congress after his experience as secretary of the environment. "Creator of the Rodízio Loses His Seat," read the headline in the *Estado de São Paulo* (8 October 1998); Feldmann had received just under 50,000 votes, well short of the 94,000 he had received in 1990 (interview with Feldmann 1991). He was not alone: more than half the members of São Paulo's delegation who ran also were not reelected to office, but he was one of the few whose defeat was so widely noted. Feldmann's largely middle-class constituency, drivers of cars, presumably turned away from him because of the rodízio, although no reliable studies have tracked this. Certainly everyone around Feldmann assumed that this was so, with his frequent critic from Greenpeace remarking wryly that Feldmann had "acted more like an environmentalist than a politician, and got hurt for it" (interview with Rodrigues 2000). Governor Covas had insisted through the election that he would maintain the rodízio if reelected against three opponents who pledged to cancel it (*Folha de São Paulo*, 24 September 1998), but his new environmental secretary announced the end of the environmental rodízio almost immediately and the issue has not been raised since.[12] Presumably the secretary's opposition was strong, since giving up the rodízio meant giving up the nearly $900,000 in fines that it had raised in 1998 (*Diário Oficial*, 5 June 1999).

Was the rodízio environmentally effective? High adherence to the rodízio did mean that there were absolute reductions in emissions in 1997 and 1998, according to CETESB. In both years over 93 percent of cars complied with the program, and 25,000 (1997) and 55,000 (1998) tons of carbon monoxide were not emitted as a result (SMA / CETESB 1998, 1; CETESB 1998b, 11,13). The overall conclusion for 1998 was that carbon monoxide emissions were down 19 percent with the rodízio, almost exactly equivalent to the 20 percent of cars that should have been off the road (CETESB 1998b, 13). For the first time in ten years there were no air quality alerts for carbon monoxide. Yet the authors of the report are hesitant to conclude that the rodízio alone was responsible, noting that meteorological conditions were very favorable and that the fleet had been renewed with new car purchases (CETESB 1998b, 20). In conversations with CETESB's air pollution technicians, none have criticized the rodízio, but no one has demanded its return either.

Environmental Justice: Urban Issues in the New Millennium

Recent city and state governments have not made many innovations in urban environmental politics in São Paulo and elsewhere in Brazil. Declining budgets through the 1990s and since have meant little scope for new state initiatives and little ability to address continued unevenness in environmental protections across Brazil. Nonstate actors have filled some of the resulting gap by creating a new Environmental Justice Network (Rede de Justiça Ambiental) in 2001. The network was inspired in part by the example of North American environmental justice activists and scholars, some of whom were present at the founding meeting in 2001, but the environmental justice idea is also consistent with the history of environmental mobilization in Brazil. This is one of the clearest examples of international diffusion of a specific environmental organizing concept to Brazil since the original conservation movements of the 1950s.[13] Few ideas from elsewhere have resonated so well in Brazil and been so deliberately imported.

While the language of environmental justice is recent in Brazil, the underlying idea—which causally links degradation of the physical environment to social inequalities and distortions created by the Brazilian model of development—is not. There is an especially clear connection to past concepts like socio-environmentalism and to claims made in the 1980s that political and economic exclusion were causes of environmental problems. While the concepts are broadly similar, the discourse of socio-environmentalism was still mainly employed by a set of self-described environmental organizations as a language with which to reach out to other movements.

The creators of the Environmental Justice Network saw environmental justice as a unifying concept that could bring together groups isolated in their struggles against globalization and economic restructuring, some environmentalist and some not (Rede de Justiça Ambiental, Manifesto, 2001). The concept has worked particularly well for trade unions, for many of which even socio-environmentalism was too far a conceptual stretch, and one that seemed to threaten jobs (interview with Malerba 2005). These concerns echoed those of some participants in earlier movements to clean up the industries of Cubatão. The peak trade union organization CUT had been active on environmental issues even before it took part in leading the preparations for the Global Forum of the Earth Summit (Marcelos Neto 2003), but rank-and-file unions had lagged behind on this issue. Most of them find environmental justice a good way to express one of their demands, without limiting their ability to demand justice in other areas.

The new network defines environmental justice and injustice in Brazil broadly: "By environmental injustice, we mean the mechanism by which societies which are economically and socially unequal place the heaviest burden of environmental degradation on low income populations, on racially discriminated groups, on traditional indigenous ethnicities, on working class neighborhoods, on marginalized and vulnerable populations" (Rede de Justiça Ambiental 2001). This definition takes environmental justice and injustice beyond the urban "brown" issues which have characterized much of the North American movement. The displacement of rural communities by large economic projects is a central concern of the Brazilian environmental justice movement (Leroy 2004). Violence against smallholders and organizers in rural and forest areas, like the murder of Dorothy Stang which opens this book, are clearly examples of environmental injustice. In another example, the landless movement uses the language of environmental justice in its critique of genetically modified organisms.

However, participants at the first conference of the network (the first formal meeting since its founding) chose to emphasize the urban character of environmental justice. Identifying a social dimension to urban environmentalism in Brazil was the real innovation, expressing a desire "to combat the idea that environmental issues are only related to nature" (Rede de Justiça Ambiental 2004, 11–12). In practice, the most active groups in the Environmental Justice Network are groups of *"expostos,"* or workers who were exposed to environmental hazards, and their work gives the network an urban cast (interview with Malerba 2005). Other activist groups include the Association against Persistent Organic Pollutants (ACPO)[14] and the Brazilian Association of Those Exposed to Asbestos (ABRAE), as well as the Movement of Those Affected by Dams (MAB). MAB and its precursor organization were part of the preparations for the Earth Summit of 1992 discussed in chapter 3, but ACPO, ABRAE, and many of the non-environmental groups which participate in the Environmental Justice Network have been formed since the conference. The first working group of the new network reflects the interests of the *expostos*; named Redeban, it intends to mobilize support for banning numerous products that are environmentally risky, from chemicals to genetically modified organisms. In mid-2005 the network was initiating contacts with Brazil's black movement, in hopes of establishing a second working group that could combat Brazil's undeniable racial inequalities. These inequalities have an environmental dimension as well as their more widely known social and economic ones (Oliveira 2003; Paixão 2003).

Many of the activities that the Environmental Justice Network opposes are illegal and represent failures in the licensing, monitoring, and production processes. To some extent they can be addressed through the legal system. The "justice" of environmental justice is less a legal concept than a political and moral one, however. From the standpoint of most laws, exactly who is negatively affected by environmental degradation is immaterial; from the standpoint of environmental justice this legal indifference masks the fact that the victims, the exposed, the dislocated come consistently from those parts of the population least equipped to prevent the degradation or mitigate its effects. For Jean-Pierre Leroy of the NGO FASE, one of the founders and coordinators of the Environmental Justice Network in Brazil, the most important legal foundation for environmental justice is found not in legislation but in the 1988 Constitution and the documents produced by the Earth Summit of 1992, which speak of human rights and citizenship (Leroy 2004, 2–4). The Constitution is rich in guarantees of both individual and collective rights to a sound environment for *everyone*.

If equal rights to a sound environment are the core idea of environmental justice, securing those rights requires making a priority of the environmental conditions experienced by the least privileged members of society. With Brazil continuously ranked as one of the most unequal countries in the world (in terms of both income distribution and access to resources), the network also asserts in its founding manifesto that this kind of injustice is central to the logic of the nation's economy (see also Acselrad, Herculano, and Pádua eds. 2003a). Although economic inequality is the root of much of the environmental injustice, political and social inequality lock it in place. The Environmental Justice Network aims to name this inequity and combat it. In pursuit of this goal, the network has issued strong statements like one condemning the Lula administration's economic policies (Rede de Justiça Ambiental 2004, 17–18), but the network really takes shape in numerous more concrete campaigns.

One of its most innovative campaigns opposes the geographic relocation of environmental hazards and risk from more to less regulated areas. There are very different levels of environmental protection in Brazil's twenty-seven states and its federal district, which we have mentioned repeatedly. The strength of laws, environmental agencies, and social movements varies dramatically from one part of Brazil to another, and the three tend to go together. Led by ACPO and the working group Redeban, the Environmental Justice Network works explicitly against the logic of "not in my backyard" (Malerba n.d.) by protecting everyone's backyard.

We see this in ACPO's activities, which take us back to Cubatão and demonstrate the very partial nature of the earlier "solution" of Cubatão's environmental problems. ACPO was founded in 1994 by workers at a chemical plant in Cubatão owned by the French multinational company Rhodia (Gomes 2003). Some of them had originally worked for a related Cubatão company, Clorogil, which public pressure forced to close in 1978, after workers' deaths from probable workplace contamination. (During Feldmann's term as state environmental secretary, the man who had been regional manager of CETESB at the time was finally removed for his inaction, some fifteen years later.) In 1993 Rhodia also closed its factory, but a lawsuit brought by the Ministério Público obliged it to guarantee jobs and medical treatment to the workers contaminated in the two facilities. When their union agreed to allow Rhodia to back out of this commitment, some of the workers formed ACPO and initiated a new set of lawsuits through the Ministério Público. While ACPO won many of these cases, members of the group have been frustrated by the slow translation of their judicial victories into compensation for their health problems. In arguments to the Ministério Público and the International Labour Organization in the mid-1990s, ACPO itself drew a connection between environmental contamination and the infringement of rights (Malerba n.d.). This understanding eventually linked ACPO to the environmental justice idea and extended to other areas of activity, including the nontransference of environmental hazards and risk.

ACPO tried to fight chemical contamination through the regulatory process but could only achieve what it considered small changes. However, it has managed four times to prevent the stockpiling and incineration of toxic wastes outside São Paulo, where environmental standards were lower, achieving through networking what it could not do through legislation (Malerba n.d., 7). This became a strategy of the Environmental Justice Network, known as the Permanent Campaign against the Transfer of Toxic Wastes between States of the Federation. This campaign uses the information and activism of groups in more densely organized parts of the country to block companies trying to escape their reach. The network looks for local groups in the targeted areas, who themselves provide crucial information and activism—and sometimes become part of the network. A similar campaign prevented the firm Aracruz Celulose from setting up eucalyptus plantations in Rio de Janeiro after they were blocked in indigenous lands of neighboring Espírito Santo (interview with Malerba 2005). Recognizing that Brazil itself is an area of lower environmental regulation compared to some parts of the world, ACPO has also led the way in joining international networks and negotiations in hopes of preventing similar transfers of environmental hazards and risks to Brazil. ACPO participates in global networks aiming to ban or limit asbestos, mercury, persistent organic pollutants, incineration, and water fluoridation, and accompanied the negotiations of the Stockholm Convention on Persistent Organic Pollutants (interview with Saussuna 2005).

The very existence of a functioning network helps to prevent some of the political and economic inequalities that the Environmental Justice Network has identified. The network is now better able to follow and influence the scale of global capital choices, as businesses move from other parts of the world to Brazil or move around in Brazil in search of greater legal and political freedom to degrade the environment. It can identify connections between the mercury contamination caused by gold miners in the Amazon and that caused by multinational corporations in São Paulo and link the two sets of victims. Politically the national and international networks can join forces to support specific mobilizations in particular places. In addition, the Environmental Justice Network has taken some steps toward establishing a pool of experts who can advise members of the network and others on legal matters and carry out health assessments to counter the experts of businesses. Funding from the Ford Foundation and the Heinrich Böll Foundation helps to support these activities. Overall, the networking helps to redress some of the imbalance in the distribution of power and, perhaps, the distribution of environmental harms and resources. At the same time, the Environmental Justice Network has succeeded primarily in blocking activities and is often a step behind environmental degraders. In the absence of effective state environmental leadership, the result is a holding pattern.

Air pollution policy in São Paulo illustrates the potential and limitations of enabling networks, consistent with Keck's findings on water pollution politics in the São Paulo metropolitan area (2001). Throughout the 1980s and the 1990s, state actors and environmentalists often worked together to push their shared agendas forward. Organized environmental movements, grassroots groups, and public opinion all provided state agencies with crucial political support and an impetus for environmental projects, especially to influence other governmental actors and public opinion. Governmental actors, especially in the state-level environmental agencies, coordinated the wide array of environmental actors and often created meaningful opportunities for participation by nonstate actors. "Government" could include people who were historically or even concurrently environmental activists, with Feldmann a notable example.

The environmental networks were rarely able to secure their aims completely, with the vaguely defined and competing competencies at three levels of government a key problem. In Feldmann's critical conclusion, "When the union [the federal level] is necessary, it's missing; when it is not necessary, the union is there" (interview with Feldmann 1999). The conflicts between state and municipal levels also complicated good air pollution policy, as did the competing responsibilities for environmental, transport, and industrial policy.

The environmental stakes became ever more complex over three decades of pollution control in São Paulo. The first generation of policies required confronting very powerful economic actors, including some under the rubric of the national state itself, but comparatively few of them. Democratization, even during the transitional phase, gave state-level agencies and their societal allies special opportunities to take on their opponents, helped by critically timed disasters that raised the public profile of air pollution. The licensing process engaged a similar set of actors, but now in routinized interactions out of the public eye. The technical skills and modernist vision that made São Paulo an early leader in Brazilian environmental politics were less adequate for dealing with air pollution from 5.5 million cars circulating in the metropolitan region. While fuel and engine standards provided certain entry points for controlling pollution, they did not resolve the broader problems of access to transportation, urban safety, and their consequences for increasing use of automobiles in the city. Such intractable issues are even more pronounced when environmental justice enters the equation. The rationale for organizing around environmental justice is that legal and technical solutions to environmental problems have not considered how unequally these problems and the cost of solving them are distributed. Environmental justice claims are a final reminder that environmental politics is inevitably about politics itself.

Conclusion

One of the advantages of studying environmental politics over an extended period is the perspective that one gains on nonlinear change. In the social sciences we tend to look backward in time, attempting to identify the causal forces and mechanisms that produced a phenomenon in the present, or a framework that makes sense of a pattern we can identify. We seek ordered sequences. Recent work on path dependence in the historical institutionalist tradition (Pierson 2004; Mahoney 2000; Thelen 2003) illustrates this tendency well: something is said to be path dependent if upon being launched by a contingent event or set of events, self-reinforcing mechanisms are set in motion that "lock in" the phenomenon, making it particularly difficult to change. The writing of the economist Brian Arthur (1990; 1999) on positive feedback and increasing returns, one of the inspirations for Pierson's approach, is part of a broader body of theory cutting across the natural and social sciences, sometimes known as complexity theory, which explores the functioning of complex adaptive systems. Among other things, complexity theory points to the impact of random or contingent events in confounding the expectations that linear or equilibrium models might predict. Complexity does not mean utter randomness, but merely that a system has regularities difficult to describe (Axelrod and Cohen 1999, 16).

In the course of our research we repeatedly encountered situations that surprised us. Expecting to trace a process of cumulative development of institutions and organizations that responded to both internal and external understandings of environmental issues and problems, we found instead a process marked by discontinuities, contingencies, dead ends, and sudden surges of opportunity. Sometimes there were advances despite seemingly impossible opposition, and sometimes procedures that seemed well established were eliminated overnight. Not only the early stages, but also the *maintenance* of institutional processes frequently required the continuing active agency of actors outside the institutions themselves. In concluding our book, we try to trace out the continuous and discontinuous elements of Brazilian environmental politics over the last thirty-five years or so, and consider the implications of these findings for social science scholarship.

Brazil combines the environmental problems of an industrialized country with all the environmental choices and challenges facing a developing one. Through cycles of economic growth and decline, Brazil achieved one of the world's largest gross national products, while maintaining one of the world's most unequal societies. The persistence of widespread poverty reinforces the argument that environmental protection needs to be reconciled with the needs of development, for the population at large and for most Brazilian environmentalists as well. Since the very beginning of environmental debates in Brazil, this has been a persistent and challenging refrain. From a discursive standpoint, to be critical of economic development per se would be to speak unintelligibly. Instead, Brazilians commenting on the environment have variously insisted that development needs must come first (at the Stockholm conference in 1972); crafted homegrown balancing terms like the Green Party's *poliséria* (pollution plus misery) and the environmental movement's socio-environmentalism; created the concept and reality of the extractive reserve; articulated every possible position on the balance in the Amazon; and borrowed the environmental justice frame from international actors. The many variations on this theme show both how difficult the combination is in practice and how important it has been to Brazilians to work it to some point of resolution.

Another persistent problem for the analyst of Brazil's environmental politics is the mutability of institutions, with frequent changes in institutional design and procedures. As new chief executives (at federal or state levels) seek to put their stamp on government, they move environmental agencies from one jurisdiction to another, change their attributions, create new departments, and eliminate others. When frequent reshuffling occurs, it becomes almost as surprising when there is policy continuity as when there is not. Sometimes institutional memory is maintained by dedicated public officials who manage to remain involved in a particular area; sometimes it comes from external monitoring, including agreements with foreign lenders or donors. However, even in institutions like CETESB, where considerable technical expertise has built up over time, politically motivated transfers of employees from one part of the agency to another can cut off the head of a policy process—or restore it. Moreover, even when policies are in place, the mechanisms that ensure their enforcement may not be. Therefore we find that it makes little sense to study institutional developments purely in terms of formal structures, policies, and procedures. Instead, studying the role of institutions regarding a particular issue requires identifying the constellation of linkages that sustain or undermine a particular policy or set of policies, both inside and outside government, and the mechanisms through which this process occurs.

Environmental policy is by its very nature relational—whether it is reacting to past action or inaction, mitigating existing or potential damage, or regulating activities whose unregulated operation would be harmful. At its best, environmental considerations would be part and parcel of all decision making, what the environmental minister Silva called transversality in 2003, but that is something which neither she nor anyone else has been able to achieve. In any case it makes no sense to study the actions of those seeking to repair environmental damage without understanding what sustains the actions of those causing it in the first place, and there has been much scholarly work on Brazilian environmental issues in this vein. There has been much less that examines the political and institutional processes involved in light of the Brazilian political process as a whole.

Many expected democratization in the 1980s to improve the conditions under which environmentalists could persuade or pressure politicians to promote better environmental legislation and policy, and in many respects it did. The end of press censorship made it possible for both print and electronic media to run news stories critical of the government. Cubatão had been a national security area under the military, unable to elect its own mayor. Early information about medical deformities there was leaked out, and enterprising members of the state assembly took up the cause of the Cubatão victims even as they supported full democratization of the regime. The Montoro government, elected in 1982, responded quickly to the situation in Cubatão, in contrast to previous military appointees in the governorship. Democratization also gave a rationale for including citizens in decision making and giving them new access to the judiciary. But it left many gaps.

Environmentalists were surprised at their electoral weakness the first time they mounted a significant effort at the polls in 1986. They have never managed since to make effective use of the party system, with the Green Party remaining one of Brazil's weakest (although succeeding in a limited way simply by surviving) and parties like the PT, PSDB, and early PMDB offering only sporadic support for environmental efforts. Still, the combination of creative strategies and a sustained voluntary effort helped Deputy Fábio Feldmann to get many important environmental goals and values written into the Constitution of 1988. Like so many other goals set forth in that massive document, these were little more than declarations of intent. Democracy was no magic bullet, and neither a constitutional provision nor strong legislation carried with it any guarantees of being implemented. Making law real required a much longer—and protracted—struggle, and it required state building (Keck and Abers 2006), which for environmentalists had to be a participatory process. Without their active involvement, they were convinced, the politicians and bureaucrats in the state and federal capitals would simply compromise away any gains that had been made.

To understand domestic politics in Brazil, one must be aware of important structural features of the system that form a backdrop for our stories—presidentialism and federalism; multiple, mainly non-programmatic parties; a bicameral federal legislature; proportional representation with high district magnitude (each state is a single district with many seats) and a nominal vote (voters vote directly for their representatives); and so on. However, together with a grasp of the formal structures and limits of the system, we must also hold on to a recognition that many formally defined limits and jurisdictions are decidedly fuzzy in practice. To institutionalists schooled in the merits of Weberian legal-rational authority, these irregularities are deeply troubling. And yet, as Chalmers pointed out in an article (1977) which we discuss extensively in the Introduction, the institutionalization of authority structures in a context of huge social disparities might well close even more doors to political and social change.

Whether one finds tolerable or intolerable the boundary blurring, pliable procedures, and many redundancies (as well as absences) in Brazilian institutional architecture, they are familiar attributes of institutions with which Brazilians work every day. The strategies of actors cannot be directed at institutions as they are imagined to exist in Western Europe and North

America, but rather as they operate in Brazil. In other words, the actors must learn to take advantage of the multiple levels at which issues can be addressed, and of the combination of formal and informal interactions necessary to sustain attention to an issue. They must be prepared to notice and take advantage of dramatic events that produce sudden disruptions in the status quo, opening a path around the usual obstacles to accomplishing goals. A year before Chico Mendes was murdered, SEMA had already approved a proposal to make extractive reserves a new kind of conservation unit; after he died, rubber tappers and their supporters were prepared to move quickly to propose areas for reserves, and they did so successfully. Similarly, when Dorothy Stang was murdered in Pará, the Environment Ministry had already prepared decrees that President Lula issued almost immediately, creating conservation units in Southern Pará alongside what was to become a major highway through the region, BR-163.

Of course contingent events can deflect attention from environmental issues as well. A good example was the rapid dissipation of the energy and attention generated by Eco-92 in the ensuing heat of the hearings concerning the impeachment of President Collor. Nonetheless, the preparations for the Earth Summit were a good example of a process with externally imposed limits and deadlines in relation to which Brazilian groups could (up to a point) organize themselves with some degree of certainty. Although the external limits certainly produced frustrations, as did the Brazilian government's failure to support the NGO proceedings as expected, they also structured a remarkable two-year process of political learning that transformed the environmental movement, the NGO sector, and the ways that both related to their foreign counterparts.

The belief in the power of participation to monitor and influence state behavior produced councils devoted to an ever-wider range of topics over the next two decades. Unlike many other countries where decentralization occurred during this period, Brazil did not decentralize at the behest of external actors (Melo and Rezende 2004). Environmental councils were established at the national and some state levels even before the end of military government. The Constitution of 1988 mandated many more, often at the municipal level, and this still preceded the beginning of Brazil's neoliberal transition. In 2001 more than 35,000 councils had been set up—on topics including education, social assistance, rights of children and adolescents, employment, tourism, transportation, and urban development—and 1,391 municipalities had at least five of them (Melo and Rezende 2004, 44). In some of the larger environmental organizations, which generally took on the job of participating in these councils (a job that required the ability to travel and to receive and process numerous documents—often at the last minute—that were expected to be the subject of deliberations), many began to complain that participation in councils was competing for time with environmental activities, as we saw in chapters 1 and 5. Nonetheless, they continued to see participation as an essential way of monitoring the expected malfunctioning of political institutions.

Participation in councils is not the only way that environmentalists have attempted to change state behavior; a more direct way has been to enter the state itself. In some cases this has occurred as newly elected governments have named prominent environmentalists to positions in the environmental bureaucracy (at the federal, state, or municipal level). The cases of Fábio Feldmann, Mary Allegretti, João Paulo Capobianco, and Rogério Rocco are only some of the better-known among the thousands who have made environmentalism their life work. Young environmentalists have become professionals, entering the civil service, state technical positions, or the Ministério Público, among others. At the same time, most of them maintain close relationships with colleagues and friends in the nonstate sector and intermittently return to it.

The complex interactions among formal and informal processes, between actors in the state and those in civil society organizations, and among people working in different parts of the state bureaucracy are the basis for what we have described here as networks of individuals. These are sometimes latent but are capable of being activated to engage in blocking activities (combining protest outside the state with resistance inside it—leaking information about potential or actual harms, foot dragging, and the like) or enabling ones (sometimes also combining external protest with internal response, or collaborating in getting procedures or policies changed or implemented). Such networks, cultivated and valued by their participants, constitute an adaptation to the difficulty of producing institutional change under normal circumstances and the need to be prepared to respond quickly to contingencies. We are accustomed to thinking of purposive networks bringing pressure on the state as tools of the powerful, or as channels for corruption, and so they may also be. Where the alternative to the agility of networks is not rule-governed institutions but rather institutions dominated by those who benefit from the status quo, building social capital among those seeking constructive change appears much the better alternative, at least in the short run. Still, however creative the solutions encountered, in the long run this very labor-intensive process of enacting policy must be transformed into public policy that can become self-sustaining, grounded as far as possible in a new common sense.

Throughout this book, we have tried to show environmental politics as embedded within national politics more generally. But although we have focused on the domestic side, transnational actors (including both foreigners and Brazilians adept at navigating transnationally) appear frequently.

At the outset, we identified a series of mechanisms characteristic of the interaction between domestic and international actors: diffusion, persuasion, leverage, payoff, and coercion. Although we found examples of all of these, their operation was never as straightforward as it appeared from the outside. Productive interactions always involved a two-way process, in which external influences were negotiated, transformed, assimilated, or fought out by domestic actors.

One of the best recent examples of diffusion was the choice of some Brazilian activists to bring the discourse of environmental justice from the United States in their efforts to bridge the twin aims of environmental protection and economic development. Their motive was the perceived failure of socio-environmentalism to provide an adequate frame—or causal story—to build new coalitions for environmental change, especially in urban areas. Earlier, diffusion certainly took place when the Brazilian military government came home from the Stockholm conference—where it had overtly resisted efforts to persuade it to adopt an environmental agenda—and created the national environmental agency SEMA. However, the agency created was only a shell, requiring the accumulation of domestic experience to give it substance. Efforts at international persuasion were clearest when Brazilians engaged outsiders in international negotiations. The never-quite-resolved debates about just how Brazilians would host the United Nations Conference on Environment and Development in 1992 illustrate the levels of efforts to persuade more than they illustrate persuasion.

The well-known stories of leverage through transnational alliances illustrate the crucial role of both blocking and enabling networks in the Amazon in particular. Their success depended on robust domestic alliances; over time, however, instances of successful leverage often dissipated in the face of the persistent sociopolitical patterns that have proved to be beyond the ability of any actor or policy to finally control. Some of their most lasting effects were in the ways foreign alliance partners modeled particular organizational styles (diffusion), or required them in exchange for project funding (payoffs). Transnational alliances, and thus leverage, were also notable for their absence from a number of areas: they were not a major part of the urban mobilizations described in chapter 5 or the self-organizing processes of the 1980s.

Still, payoffs in the form of material incentives are less central to these stories than many casual observers would assume, at least outside of Amazônia. Domestic actors fund most of the national state's environmental protection activities, through taxes, fines, and fees—the foreign share is around 13 percent. Similarly, large majorities of Brazilian environmental organizations do not receive foreign money at all. The line between payoffs and coercion is a fine one, and a number of Brazilian actors have preferred not to cross it, unwilling to risk diluting their own agendas. Nonetheless, it remains constantly present in the backlash promoted by environmentalism's opponents—in the nearly permanent charges of "internationalization" of the Amazon, in congressional inquiries into the roles of organizations like Greenpeace, and even in the sense of many environmentalists that international environmentalists have overemphasized the conservation agenda in Brazil.

In conclusion, we have found that transnational and domestic actors and processes have been heavily intertwined in Brazilian environmental politics, to the point where neither can be understood without the other and often the two cannot even be distinguished. As the relevant layers stretch from the local to the global, one of the most important strategic challenges for those who want to deepen environmental protections in Brazil is to learn how to move across those levels. Central to this effort are networks of actors, working both inside and outside the state at all levels. Although blocking activities and protests against policies leading to environmental degradation continue apace, over time these environmentalists have built increasingly robust enabling networks for effecting and internalizing positive change.

Appendix: List of Interviews

Positions current as of time of interview unless otherwise stated. Names of research assistants who conducted interviews given in parentheses.

Interviews by Kathryn Hochstetler

Aveline, Carlos. President, União Protetora da Natureza (UPAN) (São Leopoldo (RS), 3 June 1990).
Barreto, Fernando. President, Associação Campineira de Ação Ecológica (Campinas, 23 May 1990).
Barros, Luis Carlos de. Oficial, Movimento Arte e Pensamento Ecológico (MAPE) (São Paulo, 15 October 1990).
Botelho, Silvio. Diretor Geral, Secretaria Estadual do Meio Ambiente, RS (Porto Alegre, 5 September 2001).
Breda, Raquel. Office of International Relations, Ministry of the Environment (Montevideo, 6 December 2001).
Capobianco, João Paulo. Director, Fundação SOS Mata Atlântica (São Paulo, 7 August 1990).
Castro, Giselda. Co-President ADGF-Amigos da Terra (Porto Alegre, 8 June 1990).
Cunha, Renato. President, GAMBÁ (Salvador, 5 September 1991).
Elisa. President, SOPREN-Manaus (Manaus, 26 August 1991).
Farias, Alcides. President, ECOA and Executive Director, Rios Vivos (Campo Grande, 14 March 2002).
Feldmann, Fábio. Federal Deputy (PSDB) (Brasília, 14 September 1991).
Feldmann, Fábio. Former State Secretary of the Environment, São Paulo (São Paulo, 10 July 1999).
Fernandes, Domingo. Secretary General, Green Party (PV) (São Paulo, 21 June 1990).
Franco, Cicero. Member PV-SP, En Nome de Amor a Natureza-RS (São Paulo, 19 July 1990).
Frank, Heinrich. Associação Canoense de Proteção ao Ambiente Natural (Canoas (RS), 6 June 1990).
Galinkin, Mauricio. Director, Fundação CEBRAC (Brasília, 9 October 2001).

Girelli, Raimundo. Lawyer for ECOA and Rios Vivos (Campo Grande, 15 March 2002).

Gregol, Giovani. Municipal Councilor (PT) (Porto Alegre, 6 June 1990).

Gualda, Regina. Former Coordinator of Social Communication and Environmental Education, Secretaria Especial do Meio Ambiente (SEMA) (Brasília, 16 September 1991).

Heckmeir, Luiz Martins. Engineer, Air Quality Division, Fundação Estadual de Engenharia do Meio Ambiente, Rio de Janeiro (Rio de Janeiro, 22 July 1999).

Iglesias, Chico. President, Associação Potigar Amigos da Natureza, and Conama representative (Natal, 2 September 1991).

Jansen. Founder, Fórum Permanente de Debates da Amazônia (Manaus, 29 August 1991).

Kettelhut, Julio Thadeu S. Manager, Institutional Development, Secretariat of Water Resources, Ministry of the Environment (Brasília, 29 July 1999).

Lanuza, Cacilda. President, Grupo Seiva (São Paulo, 26 September 1990).

Lisboa, Marijane. Former Secretary of Environmental Quality, Ministry of Environment (MMA), former Greenpeace official (São Paulo, 12 August 2005).

Luchiese Junior, Alvaro. Director, WWF-Brasil (Brasília, 11 October 2001).

Malerba, Julianna. Oficial, Projeto Brasil Sustentável e Democrático, FASE, and administrator of Environmental Justice Network (Rio de Janeiro, 5 August 2005).

Marques, Celso. President, Associação Gaúcha de Proteção ao Ambiente Natural (AGAPAN) (Porto Alegre, 7 June 1990).

Marques, Randau. Member, OIKOS, and journalist (São Paulo, 29 September 1991).

Martins, Julio Cesar Monteiro. Os Verdes, former member of the Coordenadora Política of the PV (Rio de Janeiro, 16 July 1991).

Merico. FEEC (Xaxim, SC, 22 September 1990).

Miguel. Director, Fundação SOS Amazônia (Rio Branco, 21 August 1991).

Miller, Carlos. Director, Fundação Vitória Amazônia (Manaus, 29 August 1991).

Muser Filho[, Coronel] Manoel. Member, Companha Nacional para a Defesa e pelo Desenvolvimento da Amazônia (CNDDA) (Rio de Janeiro, 17 July 1991).

Neuhaus, Esther, and Patrícia Bonilha. Executive manager and communications director, Fórum Brasileiro de ONGs e Movimentos Sociais para o Meio Ambiente e o Desenvolvimento (FBOMS) (Brasília, 9 August 2005).

Novaes, Rosani F. F. Co-President APASB (Santa Barbara d'Oeste, SP, 16 July 1990).

Pagnocchoschi, Bruno. Ministry of the Environment, member ISPN (Brasília, 2 August 1999).

Paschoal, Elbano. Director, GAMBÁ (Salvador, 4 September 1991).

Pena, José Luis. President of the PV in São Paulo (São Paulo, 31 July 1990).

Petrillo, Celso. President, Ação Ecológica, and Conama representative (São Paulo, 22 October 1990).

Queiroz, Regiane. Member, Instituto SOS Ecológico Jurídico (São Paulo, 27 June 1990).

Rocco, Rogerio. Member, APEDEMA-Rio, Movimento de Ecologia Social (Rio de Janeiro, 1 October 1990).

Rodrigues, Délcio. Campaign director, Greenpeace (São Paulo, 15 June 2000).

Rosas, Alma. Staff member, Fundação Ondazul (Salvador, 4 September 1991).

Saraiva. President, Grupo GERMEN (Salvador, 3 September 1991).

Sarmento, Jair. 1999. Director and executive secretary, Ministry of the Environment (Brasília, 4 August 1999).

Sasseroon, Pedro. Co-President, APASB (Santa Barbara d'Oeste, SP, 24 July 1990).

Saussuna, Karen. International Coordinator, Associação de Combate aos POPs (ACPO) (São Paulo, 3 August 2005).

Schäfer, Wigold. President, Associação de Preservação do Meio Ambiente do Alto Vale do Itajaí, and Conama representative (Xaxim, SC, 22 September 1990).

Schmidt, Daniel Egon. Manager, Vehicles Inspection and Programs Division, Companhia de Tecnologia de Saneamento Ambiental, São Paulo (São Paulo, 15 July 1999).

Serenza, Eli. Journalist for CETESB (São Paulo, 12 July 1999).

Sirkis, Alfredo. President of Green Party and Municipal Council member (Rio de Janeiro, 2 October 1991).

Sorrentino, Marcos. President, Sodemap, and official, Apedema (Campinas, São Paulo, 28 August 1990).

Steinbaum, Volf. Sociologist, Coordination Group of Environmental Chambers, Companhia de Tecnologia de Saneamento Ambiental, São Paulo (São Paulo, 15 July 1999).

Svirsky, Enrique. Coordinator, PROAONG, Secretaria de Estado do Meio Ambiente, São Paulo (São Paulo, 7 July 1999).

Switkes, Glenn. Latin American program director, International Rivers Network and Rios Vivos (São Paulo, 16 October 2001).

Valente, Ivan. State Deputy (PT) (São Paulo, 16 August 1991).

Vieira, Liszt. Direito e Ambiente, former PT and PV legislator (Rio de Janeiro, 4 October 1991).

Interviews by Margaret Keck (unless otherwise indicated), taped

Ab'Saber, Aziz. Professor of geography at USP (Instituto de Estudos Avancados) (São Paulo, 29 April and 1 May 1991).

Afonso, Carlos, and Herbert de Sousa ("Betinho"). IBASE (Rio de Janeiro, 25 November 1992).

Alves, Fernando Vitor de Arajuo (Denise Capelo, Diadema, SP, 12 July 1991).

Ancona, Ana Lucia. Secretariat of Planning, São Paulo City Government (Denise Capelo, 1 July 1991).

Andrade, Neuza Carvalho Santos de. President, Mobilização Ecológica do Brasil (Biorn Maybury-Lewis, Rio de Janeiro, 31 August 1990).
Avelar, Ana Maria. INDIA (Porto Velho, RO, 21 August 1992).
Avellar, Maria Vitoria Mello de, and Sérgio Ricardo de Lima. Directors, Comité Ecológico Cultural da Ilha do Governador (Biorn Maybury-Lewis, Rio de Janeiro, 5 September 1990).
Barbosa, Cristiane, Ana Paula Guimarães, Ana Luiza Guimarães, and Valeria Guedes. Movimento de Resisténcia Ecológica—MORE (Biorn Maybury-Lewis, Niteroi, RJ, 3 September 1990).
Barros, Ruy de Goes Leite de. Greenpeace (São Paulo, 8 April 1991).
Basile, Sidnei. Director, *Gazeta Mercantil* (29 April 1991).
Boischio, Ana Amelia. Professor of Biology, University of Rondônia (Porto Velho, 9 May 1991).
Born, Rubens, and Antônio Carlos de Oliveira. OIKOS (São Paulo, 18 October 1990).
Bredariol, Celso, IED (Rio de Janeiro, 24 November 1992).
Bredariol, Celso Simões. Instituto de Ecologia e Desenvolvimento (Biorn Maybury-Lewis, Rio de Janeiro, 21 September 1990).
Brito de Neto, Armando Antonio. Executive director, Pró-Rio, and president, Movimento Pró-Floresta da Tijuca (Biorn Maybury-Lewis, Rio de Janeiro, 31 August 1990).
Carvalho, Paulo. Editor and publisher of *Pindorama*, bimonthly alternative news magazine (Biorn Maybury-Lewis, Rio de Janeiro, 31 August 1990).
Casas, Vidoq. Movimento Conservacionista Teresopolitano Teresópolis (Biorn Maybury-Lewis, Rio de Janeiro, 4 September 1990).
Christie, Ruth. President, Campanha Popular em Defesa da Natureza (Biorn Maybury-Lewis, Rio de Janeiro, 16 August 1990).
Colagrossi, Fernanda. President, Associação Amigos de Petrópolis: Patrimônio, Proteção aos Animais, Defesa da Ecologia—APANDE (Biorn Maybury-Lewis, Rio de Janeiro, 14 August 1990).
Colagrossi, Fernanda. President, APANDE (Rio de Janeiro, 25 November 1992).
Costa, Hermínio Gerônimo. SATS (Denise Capelo, 11 July 1991).
Costa e Silva, Rodolfo (São Paulo, 7 August 1992).
Cruz, Ana Maria Evarîsto. Chefe da Coordenadoria, CONAMA (Brasília, 17 May 1991).
Diniz, Nilo. Formerly of APEDEMA, currently environmental editor for SOS-Mulher (São Paulo, 20 November 1992).
Dowbor, Ladislau. Secretário dos Negócios Extraordinarios, São Paulo city government (19 October 1990).
Duarte, Lúcia Maria, Grupo Ação Ecológica (Rio de Janeiro, 24 November 1992).
Duarte, Lúcia Maria. President, Grupo Ação Ecológica (Biorn Maybury-Lewis, Rio de Janeiro, 5 September 1990).
Espirito Santo, Joper Padrão do. Treasurer, Associação Brasileira de Engenharia Sanitária (ABES) (Biorn Maybury-Lewis, Rio de Janeiro, 10 August 1990).

Farias, Virgílio Alcides De. Movimento em Defesa da Vida (ABC) (Denise Capelo, Diadema, 18 July 1991) (Margaret Keck, Diadema, 23 May 1998).
Feldmann, Fábio. Federal Deputy (São Paulo, 12 April 1991).
Fereira Filho, Eneas. Engineer, CAERD (Porto Velho, RO, 26 August 1992).
Ferreira, Gabriel. Secretário Adjunto de Planejamento, State of Rondônia (Porto Velho, 8 May 1991).
Fonseca, Gelson, Jr. Diplomatic aide to President Fernando Collar (Brasília, 1 August 1990).
Forjas, Paulo, SOS-Guarapiranga (São Paulo, 19 November 1992).
Gelman, Arquiteto Marco Leão. Centro de Estudos Ambientais de Rio das Ostras (Biorn Maybury-Lewis, Rio de Janeiro, 29 August 1990).
Goldenstein, Stela. Secretary of the Environment, SMA (São Paulo, 19 May 1998).
Grechi, Dom Moacir. Bishop (Rio Branco, Acre, 19 December 1982).
Gualda, Regina Elena Crespo. Coordenadora da Area de Desenvolvimento Institucional, Programa Nacional do Meio Ambiente, IBAMA (Brasília, 30 October 1990).
Guerra, Márcia. Movimento SOS Manguezais and Grupo de Educação Ambiental da FBCN (Biorn Maybury-Lewis, Rio de Janeiro, 6 September 1990).
Guimarães Jr., Renato. Environmental lawyer, ex-promotor de justicia, first Ph.D. in environmental law at USP (Campinas, 19 April 1991).
Hathaway, David. Alternative Technologies Project (AS-PTA) (Rio de Janeiro, July 1990).
Hollyver, Nelson. Movimento Pró-Floresta da Tijuca (Biorn Maybury-Lewis, Rio de Janeiro, 20 September 1990).
LaFleur, James. Consultant (Porto Velho, RO, 12 November 1990).
Leonel, Mauro. Director, Instituto de Antropologia e Meio Ambiente (IAMA) (São Paulo, 20 November 1992).
Libánio, José Carlos. Núcleo dos Direitos Indigenos (Brasília, 31 October 1990).
Lins, Heliana Feo, and José Antonio Nunes. CETESB (São Paulo, 17 August 1990).
Lourenço, Alberto. Secretaria de Amazônia, Ministério do Meio Ambiente (Brasília, 2002).
Magalhães, Rosalvo de. President, Centro de Estudos e Conservação da Natureza, Nova Friburgo, RJ (Biorn Maybury-Lewis, Rio de Janeiro, 24 August 1990).
Magnanigni, Alceu. President, Fundação Brasileira Para a Conservação da Natureza —FBCN (Biorn Maybury-Lewis, Rio de Janeiro, 10 August 1990).
Maia, João. Delegate, CONTAG (Rio Branco, Acre, 18 December 1982).
Marques, José Carlos. Assessor, política social, CUT (São Paulo, 25 August 1990).
Marques, Randáu. Journalist, *Jornal da Tarde* (São Paulo, 3 April 1991, 7 August 1992).
Mendes, Chico, and Nilson Morão. National Rubbertappers Council (Rio Branco, Acre, 19 December 1982).
Meyer, José Fernando Pires. Architect and coordinator of Housing Department, Prefeitura Municipal de São Bernardo do Campo (Denise Capelo, São Bernardo, 22 May 1991).

Milaré, Edis. Procurador de Justiça, Coordinador das Curadorias Especializadas de Proteção ao Meio Ambiente do Estado de São Paulo (São Paulo, 10 April 1991).
Nogueira Neto, Paulo. On ADEMA (Cristina Saliba, São Paulo, 3 December 1990).
Nogueira Neto, Paulo. Secretário do Meio Ambiente, 1973–85 (São Paulo, 11 April 1991).
Oliveira, António Carlos Alves de. 4th secretaria, Assembleia Legislativa de SP, assessor do Dep. Trípoli (26 March 1991).
Oliveira, António Carlos de (Tonhão) (São Paulo, 20 November 1992).
Ott, Ary. Vice-Reitor, UNIR (Porto Velho, RO, 24 August 1992).
Pagnoccheschi, Bruno. Diretor de Planejamento e Coordenador da Política Ambiental, Secretaria de Meio Ambiente (Brasília, 1 November 1990).
Paioli, Valdemar. Associação Mundial de Ecologia (Cristina Saliba, Cotia, SP, 13 December 1990).
Pena, José Luiz. Regional president, Green Party, State of São Paulo (Cristina Saliba, São Paulo, 7 December 1990).
Queródia, Ronaldo ("Pepe"). Assessor Meio Ambiente, Prefeitura Municipal de Santo André (Denise Capelo, Santo Andre, SP, 21 May 1991).
Quirino, José Francisco. Anthropologist, USP; formerly of CESP (São Paulo, 5 August 1992).
Rocco, Rogério. Os Verdes, assessor de meio ambiente for vereador Chico Alencar, PT (Biorn Maybury-Lewis, Rio de Janeiro, 29 August 1990).
Rocco, Rogério. Os Verdes (Rio de Janeiro, May 2004).
Saldanha, Ruth Viotti. Member, Regional Executive Committee, Green Party; Chief of Scarpinatti, Mauro. Director, Espaço (São Paulo, 26 November 1992; 20 August 1990).
Silva, Claudio Barbosa da. Assembleia Permanente de Entidades em Defesa do Meio Ambiente (Biorn Maybury-Lewis, Rio de Janeiro, 24 September 1990).
Sirkis, Alfredo. City Council representative. Vereador, Partido Verde (Rio de Janeiro, 24 November 1992).
Sobral, Helena. Environmental coordenator, Prefeitura Municipal de São Paulo (São Paulo, 20 August 1990).
Torres, André. (Cristina Saliba, São Paulo, 29 September 1992).
Torres, Hércules Silva. Grupo de Defesa da Natureza (Biorn Maybury-Lewis, Rio de Janeiro, 6 September 1990).
Trondoli, Eraldo. Director, Monitoring division, Secretaria de Desenvolvimento Ambiental, State of Rondônia (Porto Velho, 10 May 1991).
Valverde, Orlando. President, Campanha Nacional de Defesa e Pelo Desenvolvimento da Amazonia (CNDDA) (Biorn Maybury-Lewis, Rio de Janeiro, 17 August 1990).
Viana, Aurélio. Director, Rede Brasil sobre Instituições Financieras Internacionais (Brasília, 26 May 1998).
Vieira, Liszt. IED (Rio de Janeiro, 24 November 1992).

Von der Weid, Jean Marc. AS-PTA (Rio de Janeiro, 24 November 1992).
Waldman, Maurício. Coordinator, Meio Ambiente, Prefeitura Municipal de São Bernardo do Campo (Denise Capelo, São Bernardo, SP, 22 May 1991).
Waldman, Mauricio (São Paulo, 23 August 1990).
Zulauf, Werner. Former president, CETESB (Montoro government); director, CETESB (government of Paulo Egydio Martins); at time of interview, environmental consultant (São Paulo, 22 April 1991).

Interviews by Margaret Keck (unless otherwise indicated), not taped

Avelar, Ana Maria (Porto Velho, RO, 11 May 1991).
Bitelli, Júlio Glinternick. Special Division of the Environment, Itamaraty (Brasília, 3 August 1990; 30 October 1990).
Butler, John. Director of Amazon Program, WWF (Washington, 17 March 1992).
Carvalho, Domingos do Carmo de. DNAEE (Brasília, 15 May 1991).
Casara, Emmanuel (Porto Velho, RO, 6 August 1990).
Casara, Emmanuel. Secretario-adjunto, SEDAM (Porto Velho, RO, 7 May 1991).
Gleckman, Harris. UN Centre on Transnational Corporations (New York, 2 October 1990).
Góes, Paulo de. SEMA (Brasília, 17 May 1991).
Goldemberg, José. Secretary of Science and Technology (Brasília, 31 October 1990).
Groenveld, Wim. Executive director, IPHAE (Porto Velho, RO, 21 November 1990).
Gross, Tony. CEDI (Washington, 20 September 1990).
Koehntopp, Dagoberto. Coordenação de Meio Ambiente, IPEA (Brasília, 30 October 1990).
Lemke, Ignacio. Director, CPT-RO, vice-head of the national CPT, and candidate for governor from the PT in Rondônia (Porto Velho, RO, 6 August 1990).
Lima, José Roberto, José da Cruz de Sousa, and Emilia. Press office of Ibama (Brasília, 29 November 1990).
Lovejoy, Thomas. Deputy-secretary, Smithsonian Institution (Washington, March 1992).
Machado, Vera. Brazilian Embassy (Washington, 19 September 1990).
Millikan, Brent. (Porto Velho, RO, 10 August 1990).
Moreira, Marcílio Marques. Brazilian ambassador to the United States (Washington, 19 September 1990).
Oliveira, Antonio Carlos de. Vitae Civilae (São Paulo, 21 May 1998).
Perdigão, Francinette. President, INDIA (Porto Velho, RO, 11 May 1991).
Rocha, Gerôncio Albuquerque. DAEE; assistant to Hugo Marques Rosa, secretary of water resources; and first executive secretary, Alto Tietê Basin Committee (São Paulo, 28 and 29 May 1998).

Sant'ana, Marie-Madeleine Mailleux. Consultant, formerly of SUDECO (Brasília, 2 November 1990).
Scarpinatti, Mauro. Espaço (São Paulo, 20 May 1998).
Silva, Maria Emilia da, Coordinadora do Polonoroeste; João Trajano dos Santos, Coordinator of Planafloro; Sebastião Ferreira Farias, subcoordinator of Polonoroeste; Astreia Alves Jordão, subcoordinator of Planafloro; Emanuel Casara; and Augusto Pinto da Silveira (UNIR member of team). Group interview done jointly with Manfred Nitsch, Department of Economics, Free University of Berlin (Porto Velho, RO, 8 August 1990).
Sobral, Helena. Prefeitura Municipal de São Paulo (São Paulo, 17 October 1990).
Svirsky, Enrique. NGO liaison, Secretaria do Meio Ambiente (São Paulo, 20 May 1999).

Notes

Introduction

1 One hectare is about 2.5 acres. Data from Ministry of the Environment at www.mma.gov.br / index.php?ido=cnuc.geral&idEstrutura=66&idMenu=2074, visited 28 April 2006.
2 As of 28 April 2006, a Google search returned 51,200 hits for Cristovam Buarque Amazonia, all of which seem to refer to this incident, reported in both *Globo* and *Correio Brasilense*. Thousands of internet users circulated the article to people they knew—we received it several times ourselves. In an article published in *Observatório da Imprensa*, an organ of commentary on the press, Jan Schoenfelder writes that "Cristovam's reply, which could have been just a poke at the power structure (um singelo tapa de luva no poder), ended up becoming a manifesto for the silent masses, who took it as an opportunity to warm up old resentments, even though we don't know or understand everything there is to know about the complexity of the Amazon's problems (Schoenfelder 2005)."

1 Building Environmental Institutions

1 The Brazilian institution is sufficiently unusual that we use its Portuguese name to signal those differences. A comparison of Brazil's Ministério Público to nine near-equivalents in other countries found that the Brazilian institution was the most independent and most nearly unique in its powers to act (study cited in McAllister 2004, 102). The Ministério Público is discussed at length in the final section near the end of this chapter.
2 In confirmation of the theme of the importance of individual appointments, the IBDF was more successful in promoting conservation while headed by Maria Tereza Jorge Padua than under other leaders. See Foresta 1991.
3 FINEP was created in 1967 (replacing the Fund for the Financing of Projects and Programs set up two years earlier) to promote the institutionalization of research and graduate-level training in science and technology. It was part of an intensive mobilization in these areas in the 1970s, seen by the military as essential to Brazilian national development.

4 The Critical Minimum Size Project, in 1979 renamed the Biological Dynamics of Forest Fragmentation Project, studied the effects of forest fragmentation on habitat; one of its goals was to determine the minimum size that a conservation unit had to be have to effectively protect the biological diversity within. See www.nmnh.si / biodiversity / bdffp.htm.
5 A number of these initiatives are discussed in more detail in chapters 4 and 5.
6 Law No. 6938 (31 August 1981) created the SISNAMA; Decree No. 88,351 (1 June 1983) regulated it; and CONAMA Resolution No. 025 (3 December 1986) established its internal bylaws. The text of these laws can be found in Republic of Brazil 1991 and CONAMA 1990.
7 While the transition process generally promoted these principles, actors in the environmental issue areas were ahead of most other sectors in implementing them. Decentralizing arrangements in most sectors were created by the 1988 Constitution, not in 1981 (Souza 1997).
8 The 1981 SISNAMA had included an agency with this name, but it was not actually created until this version in 1989, when a version of IBAMA was formed with functions different from the original.
9 See UPAN 1990. Lutzenberger had founded the environmental organization AGAPAN in 1972, often labeled the first activist ecological organization in Brazil (see chapter 2). He also wrote several books about the environment (Lutzenberger 1985, Lutzenberger 1990) and won the Alternative Nobel Prize for his environmental activism. Lutzenberger died in 2002.
10 Since none of the governments list the exact components of this functional spending line in their budget reports, this comparison is only a very rough approximation.
11 Calculated from data at consulta.tesouro.fazenda.gov.br / coffin / dotacao—vs—despesa—param.asp.
12 Svirsky 2002 is a report of these discussions. PROAONG was created in 1995 out of São Paulo's efforts to implement Agenda 21; at the time the State Environmental Secretary was Fábio Feldmann, who also appears in this book as an NGO activist and environmental congressman. The formation of PROAONG was a response to a demand by environmentalists to have access to the state government outside of CONSEMA. PROAONG is especially useful for new civil society organizations that need to be oriented to both the secretariat and other institutions of environmental governance. It has also carried out training inside the secretariat on the need to work with NGOs. PROAONG has been a model of coordination between the state and NGOs for environmental agencies in other states—and in foreign countries—as well as for other substantive agencies in the state of São Paulo. It has been capably led through several administrations by Enrique Svirsky, himself a member of the NGO Instituto Socioambiental (interview, 1999).
13 In 1995, for example, policy formulation on the Amazon was moved to a new

National Council of Legal Amazonia, with no NGO participation. This problem was somewhat alleviated when the long-time activist Mary Allegretti came to head the Amazon agency within the Ministry of the Environment, a position she occupied in both the second Cardoso and Lula administrations (see chapter 4).

14 This account of the participation of environmentalists in the Constituent Assembly's activities is largely based on a two-thousand-page report filed with the archives of the Constituent Assembly, held in the library of the Câmara dos Deputados. See Frente Nacional de Ação Ecologica na Constituinte 1988. In addition, we used the archives of Feldmann's office in São Paulo, and conducted interviews with Feldmann and other participants.

15 Ata da 14ª Reunião da Subcommissão de Saúde, Seguridade e do Meio Ambiente (Câmara dos Deputados 1988).

16 The discussion of Lutzenberger's period as secretary draws from extensive observation by both authors as participants, and from a great many informal conversations with both Brazilian and North American environmental activists during this period.

2 National Environmental Activism

1 As one of the environmentalists jailed by the military, Marques knew the limits of this respect, however.

2 It is worth noting that until very recently, most of these things were astronomically expensive in Brazil. Installation of a phone line cost upwards of $1,000, and then only after a long waiting period. The price of phone calls, especially long-distance, was also very high. Computers, before the end of the effort to protect the Brazilian informatics industry in the early 1990s, cost around twice what they did in more developed countries. All these factors made it difficult for local organizations to acquire infrastructure of their own.

3 While this was the first major confrontation between environmentalists and the state in São Paulo, there had been smaller mobilizations by environmentalists (see chapter 5).

4 We do not know of any scholarly work on the role of Rotary Clubs in the Brazilian environmental movement, despite the extremely important role they have played. Beginning in the late 1970s, Rotary Clubs provided consistent forums for debate, funded organizations, sponsored youth groups involved with conservation activities, and more. In 1978 Professor Odsio Francisco of the University of Londrina, Paraná, was one of nine members of the International Rotary Club Commission on the Environment (*Financial Times*, 23 September 1978).

5 Maria Helena Antuniassi and her associates conducted this interview between 1985 and 1987. She generously shared the undated transcript with Hochstetler for use in her research.

6 The name of this organization does not translate easily. Arrecadação can mean fund raising, and it can mean custodianship. We have chosen to leave it in the original Portuguese.
7 Primary documents from environmentalists' debates referred to in this section can be found in the archives of the Centro de Estudos Rurais e Urbanos, USP, and were collected by Maria Helena Antuniassi and her research associates. See also Antuniassi, Magdalena, and Giansanti 1989.
8 See the publication *Lista Verde*, which was published in São Paulo by the group Pensamento Ecológico (Ecological Thought) and reported on the activities of the CIEC.
9 Primary documents about the Green Party cited in this section can be found in the national archives of the PV, in Rio de Janeiro.
10 Interview with José Lutzenberger, "O Modesto Trabalho de Refazer o Mundo," *Isto É* 162 (1980), 43–44.
11 Sirkis, who served for much of the 1990s as the Green Party's president, coined the term to distinguish these former exiles from those who began their environmental activism within Brazil.
12 The Portuguese phrase is "um socialismo ecológico autogestionário." In interviews in 1990 and 1991, many Green Party activists explicitly rejected both socialism and capitalism as failed models of development that rely on many forms of oppression, and the word "socialism" does not appear in the party's founding documents (Manifesto, Statutes, Program). However, ideas like self-governing ecological socialism figured prominently in the discourse and writings of Green Party leaders when the party was being founded and were a central theme in the campaigns of 1986 in Rio de Janeiro (Partido Verde 1986).
13 This interview was conducted by Maria Helena Antuniassi and her associates between 1985 and 1987. Professor Antuniassi generously provided transcripts of the interviews to Hochstetler.
14 For a summary of this point of view see "Partido Verde Hoje?," *Sobrevivência* 1, no. 1 (June 1985), 8–9. This was the bulletin of the environmental group AGAPAN.
15 "Documento Final, Reunião Interestadual de Ecologia Política do Rio de Janeiro" (Rio de Janeiro: Mimeographed, 24 November 1985). Of the twenty-nine participants in this gathering, nineteen came into the meeting committed to the idea of a green party, with fourteen of those from Rio de Janeiro and the remainder from Santa Catarina and Minas Gerais. Other participants came from Rio de Janeiro, Rio Grande do Sul, Santa Catarina, Minas Gerais, and the Federal District.
16 The six were primarily from the South and Southeast and included Goiás, Paraná, Rio de Janeiro, Santa Catarina, and São Paulo. The four were Alagoas, Bahia, Mato Grosso, and Minas Gerais.

17 The totals were eleven from the PT, three from the PMDB, two each from the PSB and PDT, and one each from the PMN and PTB.

3 From Protest to Project

1 A survey of 234 NGOs in 1986 found that 28.8 percent of them were formally (36) or informally (30) affiliated with the Catholic Church (Landim 1998, 36).
2 Keck witnessed firsthand many of these changes in the early 1980s and knew many of the participants; parts of this section draw on her experiences during that period.
3 Interestingly, the two people cited here both worked in the Ministry of the Environment in the Lula administration after 2003, with João Paulo Capobianco spending the interim years in large environmental NGOs and Marcos Sorrentino moving into university teaching as well as continuing activism in smaller environmental groups.
4 In stark contrast, eighty-seven of the larger environmental NGOs in the United States had an average of 183,000 members each and together shared over twelve million members (Brulle 2000, 243).
5 This was a very contentious issue in the 1980s and early 1990s in Brazil and elsewhere. The argument against the swaps was that while providing comparatively small amounts of money for environmental protection, they legitimated a foreign debt (accumulated during the military government) that many environmentalists found illegitimate. See Schilling, Waldmann, and Cruz 1991 for a summary of these arguments. The sensitivity of the discussions is illustrated by an anecdote: Hochstetler was allowed to observe a national meeting of environmental organizations in May 1990 to discuss the swaps, because of a mistaken belief that she was a representative of AGAPAN. When the mistake was discovered, she was asked not to reveal the names of participants and was refused copies of the meeting documents.
6 This survey reported on the responses of a representative sample of 216 organizations out of a larger group that replied to the questionnaire, selecting them for regional representativeness in particular (Pizzi 1995, 42). The survey had been sent to 1,563 environmental NGOs.
7 It is unclear in this study what is meant by fees, as we know of many environmental organizations that raise money through sale of T-shirts and other items, and others that accept fees for mounting environmental education programs.
8 It is unclear that Lewis's study would have necessarily identified the latter kind of funding as environmental funding.
9 For details on legislative discussions see the extensive documentation on the web sites of ABONG (www.abong.prg.br) and the Third Sector Information Network RITS (www.rits.org.br). See also IBGE 2004, Introdução and Nota Técnica (no page numbers).

10 There is a lively debate about whether this is in fact true. For just a few entries in the debate see Agyeman 2005; Chapin 2004; and Guha and Martinez-Alier 1997.
11 Many tried to take it anyway, implying or even stating that environmentalists were trying to hijack a working-class hero for a green cause (Souza 1990; Hecht and Cockburn 1989).
12 Hochstetler attended this briefing.
13 This section relies heavily on our own observation of the NGO preparations for the conference. We each attended numerous international, national, and state-level meetings between 1990 and 1992.
14 Keck, who stayed at Goree's house in Rondônia during several visits there, observed the development of this process. The Earth Negotiations Bulletin is online at http://www.iisd.ca/enbvol/enb-background.htm.
15 Pro-Rio began in June 1990 with four directors: Claudio Prado of Salve a Amazônia; João Augusto Forte, a businessman in Rio; Arnoldo de Mattos, from the state Urban Development Secretariat; and Armando Antônio de Brito Neto, an entrepreneur, environmental consultant, and founder a few months earlier of an association to protect the Forest of Tijuca.
16 Account based on field notes of Biorn Maybury-Lewis, research assistant to Keck, on a meeting of APEDEMA held at FAMERJ (Federação de Associações de Moradores do Estado do Rio de Janeiro), 21 September 1990.
17 The CGT was a labor confederation promoted by center-right parties and actors and formed as a counterweight to the leftist CUT trade union, closely linked to the PT.
18 Adams was a founder of the Texas Center for Policy Studies in 1985, in Austin, and since 1996 has been executive director of CIRMA, the Center for Mesoamerican Research, in Antigua, Guatemala.
19 Information on the initial organization of Greenpeace, Brazil, comes from our own observation and from Keck's reading in the personal archives of Tani Adams, as well as extensive conversation with her, in Antigua, Guatemala, March 1999.
20 McTaggert, a former businessman and founder of Greenpeace, was the chief architect of the pragmatic and hierarchical turn of the organization during the 1980s. These "rules" are quoted without the original source in dozens of documents about the organization; we have not determined where they originally appeared.

4 Amazônia

1 Although the present discussion focuses on the nationalist dimension of debates over the Carajás project, the process and the broader programs that it encompassed were analyzed from a variety of standpoints in the magisterial

volume organized by the CNPq (National Council for Scientific and Technical Development), Brazil's national science foundation, whose list of contributors reads like a who's who in the scientific community concerned with the region (De Almeida 1986).

2 Some prominent CNDDA members were in fact members of the Brazilian Communist Party (PCB), which maintained a nationalist developmentalist position. For an example of this kind of analysis see Klein, Madeira, Dantas, and Cordeiro 1982.

3 Lúcio Flavio Pinto, commenting on reactions to the Environmental Ministry's new forest concession bill, recalled the earlier proposal and the campaign against it, suggesting that recurring, knee-jerk anti-privatization and nationalist reactions did not address the fundamental problem of how to regulate logging in an environment where the predatory extraction of hardwoods kept increasing (Pinto 2004).

4 For a recent interdisciplinary survey of the current state of scientific knowledge on a number of these issues see Wood and Porro eds. 2002.

5 The project was proposed to UNESCO by a Brazilian chemical engineer, Paulo Estevão de Berrêdo Carneiro. At the UNESCO General Conference in Paris in 1946 there were five Brazilians, of whom three were from the Oswaldo Cruz Institute, already a center of excellence for research in disease and public health. One of these, Miguel Ozório de Almeida, objected both to the dominant view at UNESCO that science should see itself transnationally and to the notion that the developing world needed international collaboration to develop excellent scientific institutions. His argument that in Latin America the relationship between scientists and the nation was important and that in Brazil this had to go through the state prefigured later objections to the institute (see Maio 2005, 117–18)

6 In this respect they resembled their counterparts in Eastern Europe and the former Soviet Union at the time.

7 Rubber tappers harvest natural latex from rubber trees by hand, by making v-shaped cuts in the bark and attaching a small cup to catch the drips.

8 This account is based on extensive interviews done in Acre in December 1982 with the rubber-tapper leader Chico Mendes, local PT organizers and candidates, the Catholic bishop Dom Moacir Grechi, and the CONTAG organizer João Maia, and on numerous informal conversations with others present at meetings and social events over a three-day visit. A fuller version of the story is in Keck 1995.

9 Carvalho (2000) chronicles the related difficulties in establishing and maintaining indigenous land rights.

10 Current information from the web site, http://www.gta.org.br.

11 Sepp Baendereck, 1920–88, was born in Yugoslavia and settled in Rio de Janeiro in 1948, where he became known in advertising and design and as a painter and

lithographer. He visited Amazônia for the first time in 1974. After a voyage on the rivers Purus, Solimões, and Negro in 1978, he and his fellow artists Frans Krajberg and Pierre Restany produced the "Manifesto of the Rio Negro," in which they endorsed an artistic approach that they called "integral naturalism" and presented at the Beaubourg Museum in Paris. Many of his paintings through the early 1980s depict scenes from the Amazon.

12 Much of the discussion of FVA comes from a long interview with the group founder Muriel Saragoussi by Philippe Lena (Lena 2002), as well as our interview with its director Carlos Miller in 1991.

13 Brazilian budgets in effect only set budget limits for executives, who may legally choose to spend any amount up to the budgeted quantity, including no money at all. See the longer discussion in chapter 1.

14 This dispute has roiled the conservation community. See the October 2000 issue of *Conservation Biology* for one forum, which includes examples from Brazil; another with more heat appeared in *WorldWatch* (Chapin 2004). Several discussions of the potential tensions between conservation and indigenous economic practices in Brazil include Freitas, Kahn, and Rivas 2004; and Schwartzman and Zimmerman 2005.

15 Annex II listing imposed much more stringent licensing requirements for mahogany sales and more consistent monitoring by the CITES secretariat.

5 Pollution Control, Sustainable Cities

1 Water pollution is a partial exception (Keck 2001).

2 As already noted, Curitiba might well capture the status of environmental leader for cities, but São Paulo does so at the state level—admittedly propelled by its especially urgent environmental problems. On CETESB's history and its claim to being "the most solid Brazilian environmental institution in the area of pollution control," see Zulauf 1994, 37. For comparisons of environmental capacity across Brazil's regions see ABEMA 1993; Ames and Keck 1997–98.

3 Made famous by the metalworkers' strikes in the late 1970s, the largest cities of the region are Santo André, São Bernaardo do Campo, and São Caetano—thus ABC.

4 The major ones were the Petrobrás refinery Presidente Bernardes; the COSIPA steel complex; seven fertilizer firms—Copebrás, IAP, Manah, Treve, Solorrico, Ultrafértil, and Pfafer; nine petrochemical firms—Estireno, Rhodia, Alba, Liquid Quimica, Liquid Carbonic, Petrocoque, Engeclor, Carbocloro, and Union Carbide; three nonmetal mineral firms—Santa Rita, Gespa, and Concretex; Santista in paper; and Titanor in metal and mechanical production (Ferreira 1993, 55–56).

5 However, as Fernandes aptly notes, criticisms were likely to prescribe greater attention to science and education as sufficient cures for the ills of the socio-

economic model; in 1983, for example, SBPC members argued that devastation in Amazonia could have been avoided if only the government had listened to the advice of its scientists—a claim that appeared to ignore all the other powerful factors pushing predatory policies in the region (Fernandes 1989, 37–38).

6 It is hard to remember today how bitter these divisions were in 1982. The PMDB accused the PT of playing into the hands of the military by insisting on creating a new party rather than staying within the opposition front. The PT in turn accused the PMDB of elitism, and sometimes even of being "flour from the same sack" as the pro-military PDS. Recriminations were deeply felt and harshly spoken. See Keck 1992, chapter 5.

7 Www.ambiente.sp.gov.br / proaong / abertura.htm, visited 22 February 2004. Numerous organizations work on water quality issues which conceivably involve industrial sources of water pollution, but it is impossible to tell from the web site whether their focus is on industrial, sewage, agricultural, or solid waste sources.

8 It is hard to trace direct causation for air pollution because of the crucial role of meteorological factors like wind or thermal inversion (interviews with Steinbaum 1999, Schmidt 1999).

9 Observation of meeting, Encontro das Entidades Ambientalistas para a Fundação da Regional 1—APEDEMA, Santa Barbara d'Oeste, São Paulo, 23 June 1990.

10 Despite this evaluation by environmentalists, it should be noted that Maluf's first municipal environmental secretary was Werner Zulauf, who was also the successful director and then president of CETESB in the 1970s and 1980s.

11 We do not pretend to be engineering experts. The concern is to give enough information to make some sense of the policies and to be able to understand judgments about environmental effectiveness. This summary is compiled from Brazilian governmental sources including CETESB 1998b; CETESB 2000; SMA 1997b; and interviews with Heckmeier, Schmidt, and Steinbaum, all 1999.

12 The municipal government launched a year-round, rush-hour-only rodízio in 1999 with traffic-related aims, but it is missing the environmental rhetoric and policy aims of the Feldmann years. Environmentalists had argued that a permanent rodízio, unlike a winter-only rodizio, would simply encourage people to buy additional vehicles. The growing size of the fleet suggests that this may be happening.

13 On the concept's and movement's origins in the United States see Bullard 1990 and Parks and Roberts 2005.

14 Without changing its acronym, ACPO, the organization has changed its name twice. It began in 1993 as the Association of those Professionally Contaminated by Organochlorates (Associação dos Contaminados Profissionalmente por Organoclorados) and then became the Association for Awareness of Occupational Prevention (Associação de Consciência à Prevenção Ocupacional) in 1998 before taking on its current name in 2000 (Malerba n.d.).

Bibliography

Government and International Organization Documents

ABEMA. 1993. *Diagnóstico Institucional dos Órgãos Estaduais de Meio Ambiente no Brasil.* Espírito Santo: Associação Brasileira de Entidades de Meio Ambiente.

Ata da 14ª Reunião da Subcommissão de Saúde, Seguridade e do Meio Ambiente. 1987. Brasília: Câmara dos Deputados.

Câmara dos Deputados. 1988. *Archivo da Constituinte.* Brasília: Câmara dos Deputados.

CEPAL / ECLAC. 2002. *Financiamiento para el Desarrollo Sostenible en América Latina y el Caribe: De Monterrey a Johannesburgo.* Santiago: Comisión Económica para América Latina y el Caribe / Economic Commission for Latin America and the Caribbean.

CETESB. 1984. *As Indústrias de Cubatão e o Meio Ambiente.* São Paulo: Companhia de Tecnologia de Saneamento Ambiental.

———. 1985. *Air Pollution Control Program and Strategies in Brazil: São Paulo and Cubatão Areas.* São Paulo: Companhia de Tecnologia de Saneamento Ambiental.

———. 1994. *Ação da Cetesb em Cubatão: Situação em Junho de 1994.* São Paulo: Companhia de Tecnologia de Saneamento Ambiental.

———. 1998a. *Efeitos da Operação Rodízio / 98 na Qualidade do Ar na Região Metropolitana de São Paulo.* São Paulo: Companhia de Tecnologia de Saneamento Ambiental.

———. 1998b. *Operação Inverno: 1998 Qualidade do Ar.* São Paulo: Companhia de Tecnologia de Saneamento Ambiental.

———. 1998c. *Sistemática de Licenciamento da CETESB.* São Paulo: Companhia de Tecnologia de Saneamento Ambiental.

———. 2000. *Relatório de Qualidade do Ar no Estado de São Paulo, 1999.* São Paulo: Companhia de Tecnologia de Saneamento Ambiental.

Collor, Fernando. 1993. "Statement by Fernando Collor, President of Brazil and President of the United Nations Conference on Environment and Development." *Report of the United Nations Conference on Environment and Development, Rio de Janeiro, 3–14 June 1992*, vol. 2, *Proceedings of the Conference*, annex II. New York: United Nations, Document A / CONF.151 / 26 / Rev. 1 (vol 2).

CONAMA. 1990. *Resoluções CONAMA 1984 à 1990*. Brasília: Conselho Nacional do Meio Ambiente.

CONSEMA. 1993. *Dez Anos de Atividades (1983–1992)*. São Paulo: Secretaria do Meio Ambiente.

Frente Nacional de Ação Ecológica na Constituinte. 1988. *Relatório de Atividades*. Brasília: Mimeographed.

Fundo Nacional do Meio Ambiente. n.d. [ca. 2000]. *Relatório de 10 Anos da Atuação*, http://www.mma.gov.br/port/fnma.

IBGE. 2004. *Fundações Privadas e Associações*, 2nd edn, http://www2.ibge.gov.br/pub/Fundações Privadas e Associações.

Lele, Uma, Virgilio Viana, Adalberto Verissimo, Stephen Vosti, Karin Perkins, and Syed Arif Husain. 2000. *Brazil: Forests in the Balance: Challenges of Conservation with Development*. Evaluation Country Case Study Series. Washington: World Bank.

LPM. 1998. *Projeto Rodízio: Pesquisa Qualitativa*. São Paulo: Companhia de Tecnologia de Saneamento Ambiental.

Office of Management and Budget. 2006. *Historical Tables, Budget of the United States Government, Fiscal Year 2006*. Washington: U.S. Government Printing Office.

PNUD. 1998. *Atlas do Desenvolvimento Humano no Brasil*. United Nations: Programa das Nações Unidas para o Desenvolvimento [CD-ROM].

Republic of Brazil. 1991. *Meio Ambiente (Legislação)*. Brasília: Senado Federal.

Secretaria de Biodiversidade e Florestas. 2004. Plano de Controle e Prevenção ao Desmatamento. Brasília: Ministério do Meio Ambiente, http://www.mma.gov.br/doc/desmatamento2003–2004.pdf.

SMA. 1996. *Operação Rodízio 95–Do Exercício à Cidadania*. São Paulo: Secretaria de Estado do Meio Ambiente.

———. 1997a. Introduction. *A Educação pelo Rodízio*. São Paulo: Secretaria de Estado do Meio Ambiente.

———. 1997b. *Por um Transporte Sustentável: Documento de Discussão Política*. São Paulo: Secretaria de Estado do Meio Ambiente.

SMA/CETESB. 1998. *Jornal do Rodízio*. São Paulo: SMA and CETESB.

SUDAM. n.d. [1974]. *Amazônia—Novo Universo / New Universe*. Brasília: SUDAM. Trans. SUDAM.

World Bank[, Operations Evaluation Department]. 1991. "The World Bank and Pollution Control in São Paulo." Washington: Mimeographed.

Other Primary Documents

AGAPAN. 1986. *Sobrevivência* I, nos. 1, 5.

Carvalho, José Cândido de Melo. 1988. Algumas Reminiscências Conservacionistas. *Boletim* FBCN.

CEDI, OIKOS, SOS. 1990. "UNCED 1992 e as ONGS brasileiras." São Paulo: Mimeographed.

CIEC. 1986a. *Lista Verde 1.*
——. 1986b. "Síntese da Plataforma Ecopolítica." São Paulo: Mimeographed.
——. 1986c. "Vote Lista Verde." São Paulo: Mimeographed.
CIEC and Pensamento Ecológico. 1986. *Lista Verde.* São Paulo: Occasional newsletter.
"Documento Final, Reunião Interestadual de Ecologia Política do Rio de Janeiro." 1985. Rio de Janeiro: Mimeographed, 24 November.
Earth Summit: The NGO Archives. 1995. Montevideo: NGONET of Third World Institute [CD-ROM].
ENEAA. 1986. "1 Encontro Nacional de Entidades Ambientalistas Autônimos." Belo Horizonte: Mimeographed.
Fórum Brasileiro de ONGS. 1997. *Brasil Século XXI: Os Caminhos da Sustentabilidade Cinco Anos Depois da Rio-92.* Rio de Janeiro: FASE.
——. 2002. *Brasil 2002: A Sustentabilidade que Queremos.* Rio de Janeiro: FASE.
Fórum de ONGs Brasileiras. 1990. Documento Final do Encontro de Brasília. Brasília: Mimeographed.
——. 1992. *Movimento Ambientalista e Desenvolvimento: Uma Visão das ONGS e dos Movimentos Sociais Brasileiros.* Rio de Janeiro: Fórum de ONGs Brasileiras / FASE.
Greenpeace. N.d. [ca. 1989]. "Propuesta para el desa rollo de Greenpeace en America Latina y el Caribe." Greenpeace International.
——. 1997. Desflorestando o Planeta: Relatório Greenpeace sobre as Madeireiras Asiaticas—em Particular da Malásia. Greenpeace International.
——. 2003. *Relatório Anual 2003.* Greenpeace International.
Instituto Socioambiental. 2005. Dados Preliminares, http://www.socioambiental.org/nsa/doc/27072005.html.
ISA. 2005. "Forest Stewardship, Illegal Possession of Land and the Rule of Law in the Amazon Region." *Socioambiental News Updates,* 9 September.
IUCN. 1970. *IUCN Bulletin.*
Jornal de Densenvolvimento Sustentável 3, no. 6, May 2005 [publication of FBOMS].
Landim, Leilah. 1998. *Sem Fins Lucrativos: As Organizações Não-Governamentais no Brasil.* Rio de Janeiro: ISER.
Landim, Leilah, and Leticia Ligneul Cotrim. 1996. *ONGS: Um Perfil: Cadastro das Filiadas à Associação Brasileira de ONGS (ABONG).* São Paulo: ABONG /ISER.
Leroy, Jean Pierre. 2004. "O Governo sob a Luz da Justiça Ambiental." *Orçamento e Política Socioambiental* 3, no. 11, 1–12.
Marques, Celso. 1986. "AGAPAN: História de uma Luta pela Vida." *Sobrevivência* 5, nos. 7–8.
Movimento de Defesa da Amazônia, São Paulo. 1979. "Princípios, Objetivos, Cárater e Organização." São Paulo: Mimeographed, April–May.
Movimento de Defesa do Meio Ambiente do Acre. 1979. "Carta Aberta em Defesa do Acre e da Amazônia." Rio Branco, Acre: Mimeographed, April.
OAB. 2006. "Supremo Tribunal Federal tem quase 200 mil processos para julgar," http://www.oab.org/noticiaPrint.asp?id=6721.

Partido Verde. 1986. PV: *Propostas de Ecologia Política.* Coleção Anima Verde, vol. 1. Rio de Janeiro: Anima.

IV Encontro Nacional das Organizações Não-Governamentais Conservacionistas. 1984. "Declaração." Declaração. Rio de Janeiro: Mimeographed.

V Encontro das Organizações Não-Governamentais Conservacionistas. 1985. "Declaração." Rio de Janeiro: Mimeographed.

Rede de Justiça Ambiental. 2001. "Manifesto de Lançamento da Rede Brasileira de Justiça Ambiental," http://www.justicaambiental.org.br.

——. 2004. "Relatório Síntese do I Encontro da Rede Brasileira de Justiça Ambiental," 25–27 November 2004. Rio de Janeiro: FASE, Mimeographed.

Rodrigues, Délcio. 1996. *Máquina Mortífera: Carros, Transporte Urbano e Poluição do Ar.* São Paulo: Greenpeace.

Sociedade Brasileira de Física. 1976. "Carta de Belo Horizonte." Belo Horizonte, MG: Mimeographed.

Souza, Herbert de. 1991. "Posiçao do IBASE Referente ao Fórum Nacional de ONGS Preparatório à UNCED-92," 24 September 1991, reproduced in Fórum de ONGS Brasileiras Preparatório para a Conferência da Sociedade Civil Sobre Meio Ambiente e Desenvolvimento—Brazil/92, *VI Encontro Nacional, São Paulo, 27–29 de Setembro de 1991, Anais.* São Paulo: Mimeographed, 27–29 September.

UPAN. 1990. "Movimento Ecológico Gaúcho Toma Posição." São Leopoldo, RS: Mimeographed.

Who Is Who at the Earth Summit, Rio de Janeiro, 1992. 1992. Waynesville, N.C.: Visionlink Educational Foundation / Tucson: Terra Christa Communications.

Books and Articles

Abers, Rebecca Neaera, and Margaret E. Keck. 2006. "Mobilizing the State: The Erratic Partner in Brazil's Participatory Water Policy." Paper presented at the 2006 Annual Meeting of the American Political Science Association, Philadelphia, 31 August–2 September.

Acselrad, Henri, Selene Herculano, and José Augusto Pádua, eds. 2003a. *Justiça Ambiental e Cidadania.* Rio de Janeiro: Relume Dumará.

——. 2003b. "Introdução: A Justiça Ambiental e a Dinâmica das Lutas Socioambientais no Brasil—uma Introdução." *Justiça Ambiental e Cidadania,* ed. H. Acselrad, S. Herculano, and J. A. Pádua. Rio de Janeiro: Relume Dumará.

Adeel, Zafar, ed. 2003. *East Asian Experience in Environmental Governance: Response in a Rapidly Developing Region.* Tokyo: United Nations University Press.

Agüero, Felipe, and Jeffrey Stark, eds. 1997. *Fault Lines of Democracy in Post-transition Latin America.* Miami: North-South Center.

Agyeman, Julian. 2005. *Sustainable Communities and the Challenge of Environmental Justice.* New York: New York University Press.

Allegretti, Mary. 2005. "Environmental Governance in Brazilian Amazonia: Conservationists, Developers, and Social Movements." Center for Latin American Studies, University of Chicago, Latin American Briefing Series Presentation, 2005, http://clas.uchicago.edu/audio/allegretti/allegretti.rm.

Almeida, Anna Luiza Ozorio de. 1992. *The Colonization of the Amazon*. Austin: versity of Texas Press.

Alves, Maria Helena Moreira. 1985. *State and Opposition in Military Brazil*. Austin: University of Texas Press.

Ames, Barry, and Margaret E. Keck. 1997–98. "The Politics of Sustainable Development: Environmental Policy Making in Four Brazilian States." *Journal of Interamerican Studies and World Affairs* 39, no. 4, 1–40.

Antuniassi, Maria Helena Rocha, Celigrácia Magdalena, and Roberto Giansanti. 1989. *O Movimento Ambientalista em São Paulo: Análise Sociológica de um Movimento Social Urbano*. São Paulo: Centro de Estudos Rurais e Urbanos, Universidade de São Paulo.

Arantes, Rogério Bastos. 2002. *Ministério Público e Política no Brasil*. São Paulo: EDUC / Sumaré / FAPESP.

Arias. Enrique Desmond. 2006. "The Myth of Personal Security: Criminal Gangs, Dispute Resolution, and Identity in Rio de Janeiro's Favelas." *Latin American Politics and Society* 48, no. 4, 53–81.

Arnt, Ricardo Azambuja, and Stephan Schwartzman. 1992. *Um Artifício Orgânico: Transição na Amazônia e Ambientalismo (1985–1990)*. Rio de Janeiro: Rocco.

Arthur, W. Brian. 1990. "Positive Feedbacks in the Economy." *Scientific American* 262, February, 92–99.

———. 1999. "Complexity and the Economy." *Science* 284, April, 107–9.

Axelrod, Robert, and Michael D. Cohen. 1999. *Harnessing Complexity: Organizational Implications of a Scientific Frontier*. New York: Free Press.

Bache, Ian, and Matthew Flinders, eds. 2004. *Multi-level Governance*. Oxford: Oxford University Press.

Barbosa, Lívia. 1992. *O Jeitinho Brasileiro*. Rio de Janeiro: Campus.

Barretto Filho, Henyo Trindade. 2003. "Meio Ambiente, 'Realpolitik,' Reforma do Estado e Ajuste Fiscal." *A Era FHC e o Governo Lula: Transição?*, ed. L. Costa. Brasília: INESC, http://www.inesc.org.br/conteudo/publicacoes/livros.

Becker, Bertha K. 1982. *Geopolítica da Amazônia: A Nova Fronteira de Recursos*. Rio de Janeiro: Zahar.

———. 1990. *Amazônia. Serie Princípios*. São Paulo: Ática.

———. 2001. "Revisão das políticas de ocupação da Amazônia: é possível identificar modelos para projetar cenários?" *Parcerias Estratégicas*, September, 12.

Benjamin, César. 1990. "Nossos Verdes Amigos." *Teoria e Debate* 12, November, 6–21.

———. 1991. "O Doutor e o Barão (Cartas)." *Teoria e Debate* 14, May, 76–78.

Berg, Marni. 2000. "Red and Green: 20 Years of Environmental Activism in Hungary." *Problems of Post-Communism* 47, no. 2, 1–11.

Bezerra, Maria do Carmo de Lima and Marlene Allan Fernandes, eds. 2000. *Cidades Sustentáveis: Subsídios à Elaboração da Agenda 21 Brasileira.* Brasília: Ministêrio do Meio Ambiente / IBAMA / Consórcio Parceria 21 IBAM-ISER-REDEH.

Binswanger, Hans P. 1991. "Brazilian Policies That Encourage Deforestation in the Amazon." *World Development* 19, no. 7, 821–29.

Bolaffi, Gabriel. 1992. "Urban Planning in Brazil: Past Experience, Current Trends." *Habitat International* 16, no. 2, 99–111.

Borelli, Dario Luis, et al. 2005. "Entrevista com Aziz Ab'Saber: problemas da Amazônia brasileira." *Estudos Avançados* 19, no. 53, 7–35.

Boschi, Renato Raul. 1987. *A Arte de Asociação: Política de Base e Democracia no Brasil.* Rio de Janeiro: Vértice / IUPERJ.

Bramble, Barbara, and Gareth Porter. 1992. "Non-governmental Organizations and the Making of US International Environmental Policy." *The International Politics of the Environment*, ed. A. Hurrell and B. Kingsbury. Oxford: Clarendon.

Branco, Samuel Murgel. 1984. *O Fenômeno Cubatão.* São Paulo: Companhia de Tecnologia de Saneamento Ambiental.

Branford, Sue, and Bernardo Kucinski with Hilary Wainwright. 2003. *Politics Transformed: Lula and the Workers' Party in Brazil.* London: Latin America Bureau.

Brilhante, Ogenis Magno. 1997. "Brazil's Alcohol Program: From an Attempt to Reduce Oil Dependence in the Seventies to the Green Arguments of the Nineties." *Journal of Environmental Planning and Management* 40, no. 4, 435–49.

Browder, John O., and Brian J. Godfrey. 1997. *Rainforest Cities: Urbanization, Development, and Globalization of the Brazilian Amazon.* New York: Columbia University Press.

Brown, David S., J. Christopher Brown, and Scott W. Desposato. 2002. "Left Turn on Green? The Unintended Consequences of International Funding for Sustainable Development in Brazil." *Comparative Political Studies* 35, no. 76, 814–39.

Brown, J. Christopher, Scott W. Desposato, and David S. Brown. 2005. "Paving the Way to Political Change: Decentralization of Development in the Brazilian Amazon." *Political Geography* 24, no. 1, 39–52.

Brulle, Robert J. 2000. *Agency, Democracy, and Nature: The U.S. Environmental Movement from a Critical Theory Perspective.* Cambridge: MIT Press.

Buaiz, Vítor. 1988. "A Ecologia contra a Barbárie." *Teoria e Debate* 4, September, 28–31.

Bullard, Robert D. 1990. *Dumping in Dixie: Race, Class and Environmental Quality.* Boulder: Westview.

Bunker, Stephen G. 1985. *Underdeveloping the Amazon: Extraction, Unequal Exchange, and the Failure of the Modern State.* Urbana: University of Illinois Press.

Caccia Bava, Silvio. 2002. "Como Tornar Nossas Cidades Mais Sustentáveis?" *Meio Ambiente Brasil: Avanços e Obstáculos Pós-Rio-92*, ed. Aspásia Camargo, João Paulo R. Capobianco, and José Antonio Puppim de Oliveira. São Paulo: Estação Liberdade.

Caldeira, Teresa P. R. 2000. *City of Walls: Crime, Segregation, and Citizenship in São Paulo.* Berkeley: University of California Press.
Campanilli, Maura, and Sérgio Leitão. 1998. Conama Facilita Licenciamento Ambiental. *Parabólicas* 37, http://www.socioambiental.org/website/parabolicas.
Campbell, Tim E. J. 1973. "The Political Meaning of Stockholm: Third World Participation in the Environment Conference Process." *Stanford Journal of International Studies* 8, 138–53.
Campos, Anna Maria. 1990. "Accountability: Quando Poderemos Traduzi-la para o Português?" *Revista de Administração Pública* 24, no. 2, 30–50.
Capobianco, João Paulo R. 1997. "Conama: Um Espaço Ameaçado de Extinção." *Debates Sócio-Ambientais* 2, no. 6, 9–11.
Carvalho, Georgia. 2000. "The Politics of Indigenous Land Rights in Brazil." *Bulletin of Latin American Research* 19, 461–78.
———. 2001. "Metallurgical Development in the Carajás Area: A Case Study of the Evolution of Environmental Policy Formation in Brazil." *Society and Natural Resources* 14, no. 2, 127–43.
Castells, Manuel. 1996. *The Rise of the Network Society.* Oxford: Blackwell.
Castro, João Augusto de Araújo. 1972. "Environment and Development: The Case of Developing Countries." *World Eco-Crisis: International Organizations in Response,* ed. D. A. Kay and E. B. Skolnikoff. Madison: University of Wisconsin Press.
CEDI, DEPSET-CUT, et al. 1988. *De Angra a Aramar: Os Militares a Caminho da Bomba.* Rio de Janeiro: CEDI.
Chalmers, Douglas A. 1977. "The Politicized State in Latin America." *Authoritarianism and Corporatism in Latin America,* ed. J. M. Malloy. Pittsburgh: University of Pittsburgh Press.
———. 1993. "Internationalized Domestic Actors in Latin América." Mimeographed, http://www.columbia.edu/chalmers/IntDomPol/htm.
Chapin, Mac. 2004. "A Challenge to Conservationists." *WorldWatch,* November–December, 17–31.
Clapp, Jennifer, and Peter Dauvergne. 2005. *Paths to a Green World: The Political Economy of the Global Environment.* Cambridge: MIT Press.
Clark, Ann Marie, Elisabeth Friedman, and Kathryn Hochstetler. 1998. "The Sovereign Limits of Global Civil Society: A Comparison of NGO Participation in UN World Conferences on the Environment, Human Rights, and Women." *World Politics* 51, no. 1, 1–35.
Coelho, Marco António. 2005. "Warwick Kerr: A Amazônia, os índios e as abelhas, Entrevista a Marco António Coelho." *Estudos Avançados* 19, no. 53, 51–69.
Cohen, Jean L. 1985. "Strategy or Identity: New Theoretical Paradigms and Contemporary Social Movements." *Social Research* 52, no. 4, 663–716.
Conca, Ken. 1997. *Manufacturing Insecurity: The Rise and Fall of Brazil's Military-Industrial Complex.* Boulder: Lynne Rienner.
Costa, Thomas Guedes da. 2001. "Brazil's SIVAM: As It Monitors the Amazon, Will It

Fulfill Its Human Security Promise?" *Environmental Change and Security Program Report* 7, 47–58.

Crenson, Matthew. 1971. *The Un-politics of Air Pollution: A Study of Non-decisionmaking in the Cities.* Baltimore: Johns Hopkins University Press.

Crespo, Samyra, and Leandro Piquet Carneiro. 1996. "O Perfil das Instituições Ambientais do Brasil." *Cadastro Nacional Ecolista de Instituições Ambientalistas,* 2nd edn, ed. Paulo Aparecido Pizzi. Curitiba: WWF.

Da Matta, Roberto. 1987. *A Casa e a Rua: Espaço, Cidadania, Mulher e Morte no Brasil.* Rio de Janeiro: Guanabara.

Darst, Robert G. 1997. "The Internationalization of Environmental Protection in the USSR and Its Successor States." *The Internationalization of Environmental Protection,* ed. M. A. Schreurs and E. C. Economy. Cambridge: Cambridge University Press.

De Almeida, José Maria Gonçalves, Jr. 1986. *Carajás: Desafio Político, Ecologia e Desenvolvimento.* São Paulo: Brasilense; Brasília: Conselho Nacional de Desenvolvimento Científico e Tecnológico.

Dean, Warren. 1987. *Brazil and the Struggle for Rubber: A Study in Environmental History.* Cambridge: Cambridge University Press.

——. 1989. "Preserving the Amazon Forest." Mimeographed, Department of History, New York University, 16 November.

——. 1995. *With Broadax and Firebrand: The Destruction of the Brazilian Atlantic Forest.* Berkeley: University of California Press.

De Holanda, Sergio Buarque. 1982 [1936]. *Raizes do Brasil,* 15th edn. Rio de Janeiro: Jose Olympico.

Della Cava, Ralph. 1989. "The 'People's Church,' the Vatican, and *Abertura.*" *Democratizing Brazil: Problems of Transition and Consolidation,* ed. A. Stepan. Oxford: Oxford University Press.

Desai, Uday. 1998. "Environment, Economic Growth, and Government in Developing Countries." *Ecological Policy and Politics in Developing Countries: Economic Growth, Democracy, and Environment,* ed. E. Desai. Albany: State University of New York Press.

Diniz, Eliezer Martins. 2000. *Growth, Pollution and the Kyoto Protocol: An Assessment of the Brazilian Case.* Oxford: Centre for Brazilian Studies / Banco Santos.

Dobson, Andrew. 1990. *Green Political Thought: An Introduction.* London: Unwin Hyman.

Doherty, Brian, and Marius de Geus, eds. 1996. *Democracy and Green Political Thought: Sustainability, Rights and Citizenship.* London: Routledge.

Draibe, Sônia Miriam. 1998. "A Política Brasileira de Combate à Pobreza." *O Brasil e o Mundo no Limiar do Novo Século,* vol. 2, ed. J. P. R. Velloso. Rio de Janeiro: José Olympio.

Dryzek, John. 1997. *The Politics of the Earth: Environmental Discourses.* Oxford: Oxford University Press.

Dryzek, John, and James P. Lester. 1989. "Alternative Views of the Environmental Problematique." *Environmental Politics and Policy: Theories and Evidence*, ed. J. P. Lester. Durham: Duke University Press.

Dryzek, John, David Downes, Christian Hunold, and David Schlosberg with Hans-Kristian Hernes. 2003. *Green States and Social Movements: Environmentalism in the United States, United Kingdom, Germany, and Norway*. Oxford: Oxford University Press.

Durant, Robert F., Daniel J. Fiorino, and Rosemary O'Leary, eds. 2004. *Environmental Governance Reconsidered: Challenges, Choices, and Opportunities*. Cambridge: MIT Press.

Emirbayer, Mustafa, and Jeff Goodwin. 1994. "Network Analysis, Culture, and the Problem of Agency." *American Journal of Sociology* 99, 1411–53.

Enloe, Cynthia H. 1975. *The Politics of Pollution in a Comparative Perspective*. New York: David McKay.

Evans, Peter. 1997a. "Government Action, Social Capital, and Development: Reviewing the Evidence on Synergy." *State-Society Synergy: Government and Social Capital in Development*. International and Area Studies no. 94, University of California, Berkeley.

———. 1997b. "Introduction: Development Strategies across the Public-Private Divide." *State-Society Synergy: Government and Social Capital in Development*. International and Area Studies no. 94, University of California, Berkeley.

Farias, Paulo José Leite. 1999. *Competência Federativa e Proteção Ambiental*. Porto Alegre: Sergio Antonio Fabris.

Fearnside, Philip. 2001. "Land-tenure Issues as Factors in Environmental Destruction in Brazilian Amazonia: The Case of Southern Pará." *World Development* 29, no. 8, 1361–72.

Fernandes, Ana Maria. 1990. *A Construção da Ciência no Brasil e a SBPC*. Brasília: Editora Universidade de Brasília / ANPOCS / CNPq.

Fernandes, Edesio. 1995. *Law and Urban Change in Brazil*. Brookfield, Vt.: Ashgate.

Ferreira, Leandro Valle, Eduardo Venticinque, and Samuel Almeida. 2005. "O desmatamento na Amazônia e a importância das áreas protegidas." *Estudos Avançados* 19, no. 53, 157–66.

Ferreira, Lúcia da Costa. 1993. *Os Fantasmas do Vale: Qualidade Ambiental e Cidadania*. Campinas: UNICAMP.

Fico, Carlos. 1999. *Ibase: Usina de Idéias e Cidadania*. Rio de Janeiro: Garamond.

Finnemore, Martha, and Kathryn Sikkink. 1998. "International Norm Dynamics and Political Change." *International Organization* 52, no. 4, 887–917.

Foresta, Ronald A. 1991. *Amazon Conservation in the Age of Development: The Limits of Providence*. Gainesville: University Press of Florida.

———. 1992. "Amazonia and the Politics of Geopolitics." *Geographical Review* 82, no. 2, 128–42.

Foweraker, Joe. 1981. *The Struggle for Land: A Political Economy of the Pioneer Frontier in Brazil from 1930 to the Present Day*. Cambridge: Cambridge University Press.

Frank, David John, Ann Hironaka, and Evan Schofer. 2000. "The Nation-State and the Natural Environment over the Twentieth Century." *American Sociological Review* 25, no. 1, 96–116.

Freitas, Carlos E. C., James R. Kahn, and Alexandre A. F. Rivas. 2004. "Indigenous People and Sustainable Development in Amazonas." *International Journal for Sustainable Development and World Ecology* 11, no. 3, 312–24.

Friedman, Elisabeth Jay, and Kathryn Hochstetler. 2002. "Assessing the Third Transition in Latin American Democratization: Representational Regimes and Civil Society in Argentina and Brazil." *Comparative Politics* 35, no. 1, 21–42.

Friedman, Elisabeth Jay, Kathryn Hochstetler, and Ann Marie Clark. 2005. *Sovereignty, Democracy, and Global Civil Society: State-Society Relations at UN World Conferences*. Albany: State University of New York Press.

Furriela, Rachel Biderman. 1997. "O Conselho Estadual do Meio Ambiente de São Paulo." *Debates Socio-Ambientais* 2, no. 6, 7–9.

Gabeira, Fernando. 1979. *O Que e Isso, Companheiro?* São Paulo: Companhia das Letras.

——. 1987a. *Diário da Salvação do Mundo*, 2nd edn. Rio de Janeiro: Espaço e Tempo.

——. 1987b. "A Idéia de um Partido Verde no Brasil." *Ecologia e Política no Brasil*, ed. J.A. Pádua. Rio de Janeiro: Espaço e Tempo / IUPERJ.

Garcia, Flávio. 1987. "O 'Pacote' Florestal que Ameaça a Amazonia." *Pau Brasil* 16, 27–40.

Geffray, Christian. 2002a. "Introduction: Drug Trafficking and the State." *Project Most: Globalization, Drugs and Criminalization: Final Research Report on Brazil, China, India, and Mexico*, 1–5. Paris: UNESCO, http://www.unesco.org/most/globalisation/drugs—1.htm.

——. 2002b. "Social, Economic and Political Impacts of Drug Trafficking in the State of Rondônia, in the Brazilian Amazon." *Project Most: Globalization, Drugs and Criminalization: Final Research Report on Brazil, China, India, and Mexico*, 33–47. Paris: UNESCO, http://www.unesco.org/most/globalisation/drugs—1.htm.

——. 2002c. "State, Wealth, and Criminals." *Lusotopie* 1, 83–106.

Gerhards, Jürgen, and Dieter Rucht. 1992. "Mesomobilization: Organizing and Framing in Two Protest Campaigns in West Germany." *American Journal of Sociology* 98, no. 3, 555–96.

Glasson, John, and Nemesio Neves B. Salvador. 2000. "EIA in Brazil: A Procedures-Practice Gap: A Comparative Study with Reference to the European Union, and Especially the UK." *Environmental Impact Assessment Review* 20, no. 2, 181–225.

Golbery do Couto e Silva[, General]. 1967. *Geopolítica do Brasil*. Rio de Janeiro: José Olympio.

Gomes, João Carlos. 2003. "A Maior Contaminação por POPs no Brasil: O Caso Rhodia na Baixada Santista." *Justiça Ambiental e Cidadania*, ed. H. Acselrad, S. Herculano, and J. A. Pádua. Rio de Janeiro: Relume Dumará.

Green, Duncan, ed. 1989. *The Fight for the Forest: Chico Mendes in His Own Words.* London: Latin America Bureau.

Grodzins, Morton. 2000. "The Federal System." *American Intergovernmental Relations*, ed. L. J. O'Toole Jr. Washington: CQ Press.

Grzybowski, Cándido. 1989. "Rural Workers' Movements and Democratization in Brazil." Working paper, Center for International Studies, Massachusetts Institute of Technology [version published in *Journal of Development Studies* 26, no. 4 (July 1990)].

Guerreiro, R. Saraiva. 1992. *Lembranças de um Empregado do Itamaraty.* São Paulo: Siciliano.

Guha, Ramachandra, and Juan Martinez-Alier. 1997. *Varieties of Environmentalism: Essays North and South.* London: Earthscan.

Guilherme, Maria Lúcia. 1982. "Problemas Urbanos de Cubatão e seu Caráter Social." *Projeto (Revista Brasileira de Arquitetura, Planejamento, Desenho Industrial, Construção)* 43, 15–21.

Guimarães, Renato, Jr. 1989. "Direitos e Deveres Ecológicos: Efetividade Constitucional e Subsídios do Direito Norte-Americano." Thesis, U. of São Paulo, 29 December.

Guimarães, Roberto P. 1991. *The Ecopolitics of Development in the Third World: Politics and Environment in Brazil.* Boulder: Lynne Rienner.

Hagopian, Frances. 1996. *Traditional Politics and Regime Change in Brazil.* Cambridge: Cambridge University Press.

Hall, Anthony L. 1991. *Developing Amazonia: Deforestation and Social Conflict in Brazil's Carajás Programme.* Manchester: Manchester University Press.

———. 1997. *Sustaining Amazonia: Grassroots Action for Productive Conservation.* Manchester: Manchester University Press.

Harvey, David. 1990. *The Condition of Postmodernity.* Oxford: Blackwell.

Hays, Samuel P. 1959. *Conservation and the Gospel of Efficiency: The Progressive Conservation Movement, 1890–1920.* Cambridge: Harvard University Press.

Hecht, Susanna B. 2005. "Soybeans, Development and Conservation on the Amazon Frontier." *Development and Change* 36, no. 2, 375–404.

Hecht, Susanna, and Alexander Cockburn. 1989. "Defending the Rain Forest and Its People." *Nation*, 18 September, 262, 291–92.

———. 1990. *The Fate of the Forest: Developers, Destroyers and Defenders of the Amazon.* New York: Harper Collins.

Hochstetler, Kathryn. 1997. "The Evolution of the Brazilian Environmental Movement and Its Political Roles." *The New Politics of Inequality in Latin America: Rethinking Participation and Representation*, ed. D. A. Chalmers, C. M. Vilas, K. Hite, S. B. Martin, K. Piester, and M. Segarra. Oxford: Oxford University Press.

———. 2000. "Democratizing Pressures from Below? Social Movements in New Brazilian Democracy." *Democratic Brazil: Actors, Institutions, and Processes*, ed. P. R. Kingstone and T. J. Power. Pittsburgh: University of Pittsburgh Press.

———. 2002a. "After the Boomerang: Environmental Movements and Politics in the La Plata River Basin." *Global Environmental Politics* 2, no. 4, 35–57.

———. 2002b. "Brazil." *Capacity Building in National Environmental Policies: A Comparative Study of 17 Countries*, ed. M. Jänicke and H. Weidner. Berlin: Springer.

———. 2003. "Fading Green: Environmental Politics in the Mercosur Free Trade Agreement and Beyond." *Latin American Politics and Society* 45, no. 4, 1–32.

———. 2004. "Civil Society in Lula's Brazil." Working paper CBS-57-04, Centre for Brazilian Studies, University of Oxford.

———. Forthcoming. "The Multi-level Governance of GMOs in Mercosur." *The International Politics of Genetically Modified Food*, ed. Robert Falkner. Hampshire: Palgrave Macmillan.

Holston, James, and Teresa P. R. Caldeira. 1997. "Democracy, Law, and Violence: Disjunctures of Brazilian Citizenship." *Fault Lines of Democracy in Post-transition Latin America*, ed. F. Agüero and J. Stark. Miami: North-South Center.

Hooghe, Liesbet, and Gary Marks. 2003. "Unraveling the Central State, but How? Types of Multi-level Governance." *American Political Science Review* 97, no. 2, 233–43.

Hurrell, Andrew. 1992. "Brazil and the International Politics of Amazonian Deforestation." *The International Politics of the Environment*, ed. A. Hurrell and B. Kingsbury. Oxford: Clarendon.

Jacobi, Pedro. 1999. *Cidade e Meio Ambiente: Percepções e Práticas em São Paulo*. São Paulo: Annablume.

———, ed. 1997. *Poluição do Ar em São Paulo e Resposta da Ação Pública*. São Paulo: CEDEC, Cadernos CEDEC No. 60.

Kant de Lima, Roberto. 1995. "Bureaucratic Rationality in Brazil and the United States: Criminal Justice Systems in Comparative Perspective." *The Brazilian Puzzle: Culture of the Borderlands of the Western World*, ed. D. Hess and R. Da Matta. New York: Columbia University Press.

Kasa, Sjur, and Lars Otto Næss. 2005. "Financial Crisis and State-NGO Relations: The Case of Brazilian Amazonia, 1998–2000." *Society and Natural Resources* 18, 791–804.

Keck, Margaret E. 1989. "The New Unionism in the Brazilian Transition." *Democratizing Brazil*, ed. A. Stepan. Oxford: Oxford University Press.

———. 1992. *The Worker's Party and Democratization in Brazil*. New Haven: Yale University Press.

———. 1995. "Social Equity and Environmental Politics in Brazil: Lessons from the Rubber Tappers of Acre." *Comparative Politics* 27, no. 4, 409–24.

———. 1998. "Planafloro in Rondônia: The Limits of Leverage." *The Search for Account-*

ability: The World Bank, NGOs, and Grassroots Movements, ed. J. A. Fox and L. D. Brown. Cambridge: MIT Press.

———. 2001. "'Water, Water, Everywhere, nor Any Drop to Drink': Land Use and Water Policy in São Paulo, Brazil." *Livable Cities? Urban Struggles for Livelihood and Sustainability*, ed. P. Evans. Berkeley: University of California Press.

———. 2002. "Amazonia in Environmental Politics." *Environment and Security in the Amazon Basin*, ed. J. S. Tulchin and H. A. Golding. Washington: Woodrow Wilson International Center for Scholars.

Keck, Margaret E., and Rebecca Neaera Abers. 2006. "Civil Society and State-Building in Latin America." *Latin American Studies Association Forum* 37, no. 1, 30–32.

Keck, Margaret E., and Kathryn Sikkink. 1998a. *Activists beyond Borders: Advocacy Networks in International Politics*. Ithaca: Cornell University Press.

———. 1998b. "Transnational Advocacy Networks in the Movement Society." *The Social Movement Society: Contentious Politics for a New Century*, ed. D. S. Meyer and S. Tarrow. Lanham, Md.: Rowman and Littlefield.

Kitschelt, Herbert. 1993. "The Green Phenomenon in Western Party Systems." *Environmental Politics in the International Arena: Movements, Parties, Organizations, and Policy*, ed. S. Kamienicki. Albany: SUNY Press.

Klein, Odacir, Luiz Carlos Lopes Madeira, Marcos Dantas, and Marcelo Cordeiro. 1982. *Salvar Carajás*. Porto Alegre: LPM.

Kolk, Ans. 1996. *Forests in International Environmental Politics: International Organisations, NGOs and the Brazilian Amazon*. Utrecht, Netherlands: International Books.

———. 1998. "From Conflict to Cooperation: International Policies to Protect the Brazilian Amazon." *World Development* 26, no. 8, 1481–93.

Krasner, Stephen D. 1985. *Structural Conflict: The Third World against Global Liberalism*. Berkeley: University of California Press.

Kucinski, Bernardo. 1982. "Cubatão: Uma Tragédia Ecológica." *Ciência Hoje* 1, 11–24.

Kütting, Gabriela. 1999. *Environment, Society and International Relations: Towards More Effective International Agreements*. London: Routledge.

Lamounier, Bolivar. 1989a. "*Authoritarian Brazil* Revisited: The Impact of Elections on the *Abertura*." *Democratizing Brazil: Problems of Transition and Consolidation*, ed. A. Stepan. New York: Oxford University Press.

———. 1989b. *Partidos e Utopias: O Brasil no Limiar dos Anos 90*. São Paulo: Loyola, Temas Brasileiras III.

Landim, Leilah. 1998. "'Experiência Militante': Histórias das Assim Chamadas ONGs." *Ações em Sociedade: Militância, Caridade, Assistência*, ed. L. Landim. Rio de Janeiro: ISER / NAU.

Landim, Leilah, Neide Beres, Regina List, and Lester M. Salamon. 1999. "Brazil." *Global Civil Society: Dimensions of the Nonprofit Sector*, ed. Lester K. Salamon, Helmut K. Anheier, Regina List, Stefan Toepler, S. Wojciech Sokolowski, et al. Baltimore: Johns Hopkins Center for Civil Society Studies.

Laurance, William F., Mark A. Cochrane, Scott Bergen, Philip M. Fearnside, Patricia Delamônica, Christopher Barber, Sammya D'Angelo, and Tito Fernandes. 2001. "The Future of the Brazilian Amazon." *Science* 291 (5503):436–39.

Laurence, William F., Scott Bergen, Mark A. Cochrane, Philip M. Fearnside, Patricia Delamônica, Sammya d'Angelo, Christopher Barber, and Tito Fernandes. 2005. "The Future of the Amazon." *Tropical Rainforests: Past, Present, and Future*, ed. E. Bermingham, C. W. Dick, and C. Moritz. Chicago: University of Chicago Press.

Lele, Uma, Virgilio Viana, Adalberto Verissimo, Stephen Vosti, Karin Perkins, and Syed Arif Husain. 2000. *Brazil: Forests in the Balance: Challenges of Conservation with Development*. Evaluation Country Case Study Series. Washington: World Bank, http://www.worldbank.org/html/oed.

Lemos, Maria Carmen de Mello. 1998. "The Politics of Pollution Control in Brazil: State Actors and Social Movements Cleaning Up Cubatão." *World Development* 26, no. 1, 78–87.

Lemos, Maria Carmen, and Johanna Looye. 2003. "Looking for Sustainability: Environmental Coalitions across the State-Society Divide." *Bulletin of Latin American Research* 22, no. 3, 350–70.

Lena, Philippe. 2002. "A atuação sócio-ambiental de uma ONG ambientalista na Amazônia: A Fundação Vitória Amazônica, Entrevista com Muriel Saragoussi." *Lusotropie* 1, 293–301.

Levitsky, Steven, and Maria Victoria Murillo. 2006. "Variation in Institutional Strength in Latin America: Causes and Implications." Paper presented at the International Meeting of the Latin American Studies Association, San Juan, Puerto Rico, 15–18 March 2006.

Lewis, Tammy L. 2003. "Environmental Aid: Driven by Recipient Need or Donor Interests?" *Social Science Quarterly* 84, no. 1, 144–61.

Lieders, Adriana. 2001. "A New Chapter in Brazil's Oil Industry: Opening the Market While Protecting the Environment." *Georgetown International Environmental Law Review* 13, 781–99.

Lopes, Juarez Brandão. 1996. "Obstacles to Economic Reform in Brazil." *Institutional Design in New Democracies: Eastern Europe and Latin America*, ed. A. Lijphart and C. H. Waisman. Boulder: Westview.

Lowry, William. 1992. *The Dimensions of Federalism: State Governments and Pollution Control Policies*. Durham: Duke University Press.

Lutzenberger, José. 1976. *Fim do Futuro? Manifesto Ecológico Brasileiro*. Porto Alegre: Movimento.

——. 1985. *Ecologia: Do Jardim ao Poder*. 10th edn. Porto Alegre: LPM, Coleção Universidade Livre.

——. 1990. *Gaia: O Planeta Vivo (Por um Caminho Suave)*. Porto Alegre: LPM.

Macedo, Joseli. 2004. "City Profile: Curitiba." *Cities* 21, no. 6, 537–49.

Mafra, Humberto, ed. 1995. *Desafios e Perspectivas do Movimento Ambientalista no*

Brasil, Report on a Seminar held in Brasilia, 28–30 September 1995. Brasília: Fundação Francisco.

Mahar, Dennis J. 1989. *Government Policies and Deforestation in Brazil*. Washington: World Bank.

Mahoney, James. 2000. "Path Dependence in Historical Sociology." *Theory and Society* 29, no. 4, 507–48.

Malerba, Julianna Eluze Carrera. n.d. "Novas Alianças para uma Nova Estratégia no Embate Capital-Trabalho: Uma Análise Sobre a Resistência dos Trabalhadores da Rhodia Contra a Transferência do Passivo Ambiental da Empresa." Rio de Janeiro: FASE, mimeographed.

Mainwaring, Scott. 1991. "Politicians, Parties, and Electoral Systems: Brazil in Comparative Perspective." *Comparative Politics* 26, no. 2, 21–43.

Mainwaring, Scott, and Eduardo Viola. 1984. "New Social Movements, Political Culture, and Democracy: Brazil and Argentina in the 1980s." *Telos* 61, 17–52.

Maio, Marcos Chor. 2005. "A UNESCO e o projeto de criação de um laboratório científico internacional na Amazônia." *Estudos Avançados* 19, no. 53, 115–30.

Marcelos Neto, Temístocles. 2003. "Sindicalismo e Justiça Ambiental no Brasil." *Justiça Ambiental e Cidadania*, ed. H. Acselrad, S. Herculano, and J. A. Pádua. Rio de Janeiro: Relume Dumará.

Margulis, Sérgio. 2003. *Causas do Desmatamento da Amazônia Brasileira*. Brasília: Banco Mundial.

Martínez-Lara, Javier. 1996. *Building Democracy in Brazil: The Politics of Constitutional Change, 1985–95*. New York: St. Martin's.

Martins, Luciano. 1997. Reforma da Administração Pública e cultura política no Brasil: Uma Visao Geral. Brasília: ENAP.

McAllister, Lesley Krista. 2004. *Environmental Enforcement and the Rule of Law in Brazil*. Diss., University of California, Berkeley.

———. 2005. "Judging GMOs: Judicial Application of the Precautionary Principle in Brazil." *Ecology Law Quarterly* 32, no. 1, 149–74.

Mecedo, Joseli. 2004. "City Profiles: Curitiba." *Cities* 21, no. 6, 537–49.

Medina, Cremilda. 1998. *Símbolos & Narrativas: Rodízio 97 na Cobertura Jornalística*. São Paulo: Secretaria do Meio Ambiente.

Melo, Marcus, and Flavio Rezende. 2004. "Decentralization and Governance in Brazil." *Decentralization and Democratic Governance in Latin America*, ed. J. S. Tulchin and A. Selee. Woodrow Wilson Center Reports on the Americas no. 12. Washington: Woodrow Wilson International Center for Scholars.

Mendes, Chico. 1989. *Fight for the Forest: Chico Mendes in His Own Words*. London: Latin America Bureau.

Menezes, Claudino Luiz. 1996. *Desenvolvimento Urbano e Meio Ambiente: A Experiência de Curitiba*. Campinas, SP: Papirus.

Micheles, Carlos, et al. 1989. *Cidadão Constituinte: A Saga das Emendas Populares*. Rio de Janeiro: Paz e Terra.

Middleton, Neil, Phil O'Keefe, and Sam Moyo. 1993. *Tears of the Crocodile: From Rio to Reality in the Developing World.* London: Pluto.

Midlarsky, Manus I. 1998. "Democracy and the Environment: An Empirical Assessment." *Journal of Peace Research* 35, no. 3, 341–61.

Minc, Carlos. 1985. *Como Fazer Movimento Ecológico e Defender a Natureza e as Liberdades.* Petrópolis: Vozes / IBASE.

———. 1991. "Verdes Revolucionários." *Teoria e Debate* 13, February, 55–61.

Mische, Ann. 2007. Forthcoming. *Partisan Publics: Multiple Networks and Communicative Styles in Brazilian Youth Politics.* Princeton: Princeton University Press.

Moisés, José Álvaro. 1990. *Cidadania e Participação.* São Paulo: CEDEC / Marco Zero.

Monteiro, Carlos Augusto de Figueiredo. 1981. *A Questão Ambiental no Brasil, 1960–1980.* University of São Paulo, Instituto de Geografia, IGEOG-USP Série Teses e Monografias no 43.

Moraes, Raimundo de Jesus Coelho de. 2005. "Judicialização do Licenciamento Ambiental no Brasil: Excesso ou Garantia de Participação." *Revista de Direito Ambiental* 10, no. 38, 204–37.

Moran, Emílio, and Stephen McCracken. 2004. *Ambiente e Sociedade* 7, no. 2, 11–43.

Najam, Adil. 2005. "The View from the South: Developing Countries in Global Environmental Politics." *The Global Environment: Institutions, Law, and Policy,* 2nd edn, ed. R. S. Axelrod, D. L. Downie, and N. J. Vig. Washington: CQ Press.

Nogueira, Luis, and Ana M. F. de Carvalho. 1985. "Ecologistas e Pacifistas se Organisam." *Pau-Brasil* 1, no. 6, 86.

Nylen, William R. 2003. *Participatory Democracy versus Elitist Democracy: Lessons from Brazil.* New York: Palgrave Macmillan.

O'Donnell, Guillermo. 2001. "Democracy, Law, and Comparative Politics." *Studies in Comparative International Development* 36, no. 1, 7–36.

———. 2004. "Why the Rule of Law Matters." *Journal of Democracy* 15, no. 4, 32–46.

Offe, Claus. 1985. "New Social Movements: Challenging the Boundaries of Institutional Politics." *Social Research* 52, no. 4, 817–68.

Oliveira, Marcelo E. Dias de, Burton E. Vaughan, and Edward J. Rykiel Jr. 2005. "Ethanol as Fuel: Energy, Carbon Dioxide Balances, and Ecological Footprint." *BioScience* 55, no. 7, 593–602.

Oliveira, Ney dos Santos. 2003. "Desigualdade Racial e Social: A Alocação dos Negros na Favela Morro do Estado, em Niterói (RJ)." *Justiça Ambiental e Cidadania,* ed. H. Acselrad, S. Herculano, and J. A. Pádua. Rio de Janeiro: Relume Dumará.

Ostrom, Elinor. 1997. "Crossing the Great Divide: Coproduction, Synergy, and Development." *State-Society Synergy: Government and Social Capital in Development,* ed. P. Evans. International and Area Studies no. 94, University of California, Berkeley.

Pádua, José Augusto. 1991. "O Nascimento da Política Verde no Brasil: Fatores Exógenos e Endógenos." *Ecologia e Política Mundial,* ed. H. Leis. Rio de Janeiro: FASE / Vozes / AIRI / PUC-Rio.

———. 1992. "The Birth of Green Politics in Brazil: Exogenous and Endogenous Factors." *Green Politics Two*, ed. W. Rüdig. Edinburgh: Edinburgh University Press.

———. 2002. *Um Sopro de Destruição: Pensamento Político e Crítica Ambiental no Brasil Escravista (1786–1888)*. Rio de Janeiro: Jorge Zahar.

Padua, Maria Tereza Jorge. 1986. "Algumas idéias de como ocupar a selva amazônica." *Pau Brasil* 2, no. 10, 89–92.

Paixão, Marcelo. 2003. "O Verde e o Negro: A Justiça Ambiental e a Questão Racial no Brasil." *Justiça Ambiental e Cidadania*, ed. H. Acselrad, S. Herculano, and J. A. Pádua. Rio de Janeiro: Relume Dumará.

Parks, Bradley C., and J. Timmons Roberts. 2006. "Environmental and Ecological Justice." *Palgrave Advances in International Environmental Politics*, ed. M. Betsill, K. Hochstetler, and D. Stevis. Basingstoke: Palgrave.

Pasha, Mustapha Kamal, and David L. Blaney. 1998. "Elusive Paradise: The Promise and Peril of Global Civil Society." *Alternatives* 23, no. 4, 417–50.

Pereira, Anthony W. 2000. "An Ugly Democracy? State Violence and the Rule of Law in Postauthoritarian Brazil." *Democratic Brazil: Actors, Institutions, and Processes*, ed. P. R. Kingstone and T. J. Power. Pittsburgh: University of Pittsburgh Press.

Pereira, Anthony. 2006. "Public Security, Private Interests, and Police Reform in Brazil." Paper presented at the XXVI International Conference of the International Political Science Association, San Juan, 15–18 March.

Pierre, Jon, ed. 2000. *Debating Governance*. Oxford: Oxford University Press.

Pierson, Paul. 2004. *Politics in Time: History, Institutions, and Social Analysis*. Princeton: Princeton University Press.

Pinto, Lúcio Flávio. 1982. *Carajás, o Ataque ao Coração da Amazônia*. Rio de Janeiro: Marca Zero.

———. 2004. A grande verdade do Sul Maravilha. Jornal Pessoal, 19 August, http://amazonia.org.br/compradores/opiniao/artigo–detail.cfm?id'121014.

Pizzi, Paulo. 1995. "O Perfil das ONGs Ambientalistas Brasileiras." *Desafios e Perspectivas do Movimento Ambientalista no Brasil*, ed. H. Makra. Brasília: Fundação Francisco.

Press, Daniel. 1994. *Democratic Dilemmas in the Age of Ecology: Trees and Toxics in the American West*. Durham: Duke University Press.

Procópio, Argemiro. 1992. *Amazônia: Ecologia e Degradação Social*. São Paulo: Alfa-Omega.

Putnam, Robert. 1988. "Diplomacy and Domestic Politics: The Logic of Two Level Games." *International Organization* 42, no. 3, 427–60.

Queiroz Neto, José Pereira de, Roque Monteleone Neto, and Marques. 1984. "Cubatão 1984: Não Deixa que Joguem Aqui um Pó de Cal." *Ciência Hoje* 3, 80–86.

Rabinovitch, Jonas. 1992. "Curitiba: Towards Sustainable Urban Development." *Environment and Urbanization* 4, no. 2, 62–73.

Rabinovitch, Jonas, and Josef Leitman. 1996. "Urban Planning in Curitiba." *Scientific American* 274, no. 3, 46–53.

Randall, Laura. 1992. "Petroleo, Economia y Medio Ambiente en Brasil." *Revista Mexicana de Sociología* 54, no. 2, 185–211.

Redwood, John, III. 1993. *World Bank Approaches to the Environment in Brazil: A Review of Selected Projects*. Washington: World Bank.

Reid, Walter V., Osvaldo Lucon, Suani Teixeira Coelho, and Patricia Guardabassi. 2005. *No Reason to Wait: The Benefits of Greenhouse Gas Reduction in São Paulo and California*. Menlo Park, Calif.: Hewlett Foundation.

Reis, Arthur Cézar Ferreira Reis. 1982 [1960]. *A Amazônia e a Cobiça Internacional*. 5th edn. Rio de Janeiro: Civilização Brasileira / SUFRAMA.

Roberts, J. Timmons. 1996. "Global Restructuring and the Environment in Latin America." *Latin America in the World Economy*, ed. R. P. Korzeniewicz and W. C. Smith. Westport: Greenwood.

Roberts, J. Timmons, and Nikki Demetria Thanos. 2003. *Trouble in Paradise: Globalization and Environmental Crises in Latin America*. New York: Routledge.

Rocha, Gerôncio Abuquerque. 1986. "A cobiça das mineradoras sobre as terras indígenas." *Pau Brasil* 2, no. 10, 75–84.

Rodrigues, Maria Guadalupe Moog. 2004. *Global Environmentalism and Local Politics: Transnational Advocacy Networks in Brazil, Ecuador, and India*. Albany: State University of New York Press.

Romano, Jesuino, Claudio Darwin Alonso, Carlos Ibsen Vianna Lacava, and Maria Helena R. B. Martins. 1995. *Estudo do Episódio de Emergência Ocorrido em Cubatão-Vila Parisi em 1994*. São Paulo: Companhia de Tecnologia de Saneamento Ambiental.

Rosenau, James. 2000. "Change, Complexity and Governance in Globalizing Space." *Debating Governance*, ed. J. Pierre. Oxford: Oxford University Press.

Rua, Maria das Graças, 1997. "Novas questões ou antigas preocupações? A burocracia na sociedade democrática do final do século XX." *A Sociedade Democrática no Final do Século*, ed. A. A. C. Trinidade and M. F. de Castro. Brasília: Paralelo 15.

Sadek, Maria Tereza Aina. 2004. "Poder Judiciário: Perspectivas de Reforma." *Opinião Pública* 10, no. 1, 1–62.

Samuels, David. 2003. "Fiscal Straitjacket: The Politics of Macroeconomic Reform in Brazil, 1995–2002." *Journal of Latin American Studies* 35, no. 3, 545–69.

Schilling, Paulo R., Maurício Waldman, and Paulo Davidoff C. Cruz. 1991. *Conversão da Dívida e Meio Ambiente*. São Paulo: CEDI-Global.

Schmink, Marianne, and Charles H. Wood, eds. 1992. *Contested Frontiers in Amazonia*. New York: Columbia University Press.

Schmitter, Philippe C. 1971. *Interest Conflict and Political Change in Brazil*. Stanford: Stanford University Press.

Schneider, Ben Ross. 1991. *Politics within the State: Elite Bureaucrats and Industrial Policy in Authoritarian Brazil*. Pittsburgh: University of Pittsburgh Press.

Schoenfelder, Jan. 2005. "O debate sobrevive, quatro anos depois." *Observatório da Imprensa* 5, http://observatorio.ultimosegundo.ig.com.br/artigos.asp?cod'320ENO005.

Schönenberg, Regine. 2002. "Drug Trafficking in the Brazilian Amazon." *Project Most: Globalization, Drugs and Criminalization: Final Research Report on Brazil, China, India, and Mexico*, 33-47. Paris: UNESCO, http://www.unesco.org/most/globalization/drugs—1.htm.

Schreurs, Miranda A. 2002. *Environmental Politics in Japan, Germany, and the United States*. Cambridge: Cambridge University Press.

Schultz, Alfred. 1967. *The Phenomenology of the Social World*, trans. George Walsh and Frederick Lehnert. Evanston: Northwestern University Press.

Schwartzman, Stephan. 1989. "Extractive Reserves: The Rubber Tappers' Strategy for Sustainable Use of the Amazon Rainforest." *Fragile Lands of Latin America: Strategies for Sustainable Development*, ed. J. O. Browder. Boulder: Westview.

Schwartzman, Stephan, and Barbara Zimmerman. 2005. "Conservation Alliances with Indigenous Peoples of the Amazon." *Conservation Biology* 19, no. 3, 721–27.

Scott, James C. 1985. *Weapons of the Weak: Everyday Forms of Peasant Resistance*. New Haven: Yale University Press.

Selcher, Wayne A. 1998. "The Politics of Decentralized Federalism, National Diversification, and Regionalism in Brazil." *Journal of Interamerican Studies and World Affairs* 40, no. 4, 25–50.

Shoumatoff, Alex. 1990. *The World Is Burning: Murder in the Rain Forest*. Boston: Little, Brown.

Sikkink, Kathryn. 1991. *Ideas and Institutions: Developmentalism in Brazil and Argentina*. Ithaca: Cornell University Press.

Silva, Cátia Aida. 2001. "Promotores de Justiça e Novas Formas de Atuação em Defesa de Interesses Sociais e Coletivos." *Revista Brasileira de Ciências Sociais* 16, no. 45, 127–44.

Silva, Eduardo. 1997. "Chile." *National Environmental Policies: A Comparative Study of Capacity-Building*. Berlin: Springer.

Sirkis, Alfredo. 1980. *Os Carbonários: Memórias da Guerrilha Perdida*. São Paulo: Global.

———. 1996. *Verde Carioca*. Rio de Janeiro: Record.

Skidmore, Thomas E. 1988. *The Politics of Military Rule in Brazil, 1964–1985*. New York: Oxford University Press.

Smith, William C., and Roberto Patricio Korzeniewicz. 1997. "Latin America and the Second Great Transformation." *Politics, Social Change, and Economic Restructuring in Latin America*, ed. W. C. Smith and R. P. Korzeniewicz. Miami: North-South Center.

Soares, Márcia Miranda, and Luiz Cláudio Lourenço. 2004. "A representação política dos estados na federação brasileira." *Revista Brasileira das Ciências Sociais* 19, no. 56, 113–27.

Soares-Filho, Britaldo Silveira, Daniel Curtis Nepstad, Lisa Curran, et al. 2005. "Cenários de desmatamento para a Amazônia." *Estudos Avançados* 19, no. 54, 137–52.

Souza, Celina. 1997. *Constitutional Engineering in Brazil: The Politics of Federalism and Decentralization*. New York: St. Martin's.

Souza, Hélcio de. 2002. *Orçamento e Política Ambiental* 1, no. 3. Brasília: INESC.

Souza, Márcio. 1990. *O Empate contra Chico Mendes*. São Paulo: Marco Zero.

Souza, Maria do Carmo Campello de. 1989. "The Brazilian 'New Republic': Under the 'Sword of Damocles.'" *Democratizing Brazil: Problems of Transition and Consolidation*, ed. A. Stepan. New York: Oxford University Press.

Spretnak, Charlene, and Fritjof Capra. 1986. *Green Politics: The Global Promise*. Santa Fe: Bear.

Steinberg, Paul. 2001. *Environmental Leadership in Developing Countries: Transnational Relations and Biodiversity Policy in Costa Rica and Bolivia*. Cambridge: MIT Press.

Stepan, Alfred. 1973. "The New Professionalism of Internal Warfare and Military Role Expansion." *Authoritarian Brazil: Origins, Policies, and Future*, ed. A. Stepan. New Haven: Yale University Press.

———, ed. 1989. *Democratizing Brazil: Problems of Transition and Consolidation*. Oxford: Oxford University Press.

Stevis, Dimitris, and Valerie J. Assetto. 2001. "Introduction: Problems and Solutions in the International Political Economy of the Environment." *The International Political Economy of the Environment: Critical Perspectives*. Boulder: Lynne Rienner.

Stone, Deborah A. 1989. "Causal Stories and the Formation of Policy Agendas." *Political Science Quarterly* 104, no. 2, 281–300.

Svirsky, Enrique, ed. 2002. *Entidades Ambientalistas Não Governamentais em Conselhos de Meio Ambiente*. São Paulo: Secretaria de Estado do Meio Ambiente.

Svirsky, Enrique, and João Paulo R. Capobianco, eds. 1997. *Ambientalismo no Brasil: Passado, Presente e Futuro*. São Paulo: Instituto Socioambiental / Secretaria do Meio Ambiente do Estado de São Paulo.

Symmes, Patrick. 2002. "Blood Wood." *Outside Magazine*, October, 116–26.

Szarka, Joseph. 2002. *The Shaping of Environmental Policy in France*. New York: Berghahn.

Tarrow, Sidney. 1994. *Power in Movement: Social Movements, Collective Action and Politics*. Cambridge: Cambridge University Press.

———. 2005. *The New Transnational Activism*. Cambridge: Cambridge University Press.

Tatagiba, Luciana. 2002. "Os Conselhos Gestores e a Democratização das Políticas Públicas no Brasil." *Sociedade Civil e Espaços Públicos no Brasil*, ed. E. Dagnino. São Paulo: Paz e Terra.

Thelen, Kathleen. 2003. "How Institutions Evolve: Insights from Comparative Historical Analysis." *Comparative Historical Analysis in the Social Sciences*, ed. J. Mahoney and D. Rueschemeyer. Cambridge: Cambridge University Press.

Tokar, Brian. 1997. *Earth for Sale: Reclaiming Ecology in the Age of Corporate Greenwash*. Boston: South End.

Tsing, Anna. 1997. "Transitions as Translations." *Transitions, Environment, Translations: The Meaning of Feminism in Contemporary Politics*, ed. J. Scott, C. Kaplan, and D. Keats. New York: Routledge.

———. 2005. *Friction: An Ethnography of Global Connection*. Princeton: Princeton University Press.

Valverde, Orlando. 1989. *Grande Carajás: planejamento da destruição*. Rio de Janeiro: Forense Universitária.

Viana, Gilney, Marina Silva, and Nilo Diniz, eds. 2001. *O Desafio da Sustentabilidade: Um Debate Socioambiental no Brasil*. São Paulo: Fundação Perseu Abramo.

Vianna, Aurélio, Jr. 1998. *A Estratégia dos Bancos Multilaterais para o Brasil: Análise Crítica e Documentos Inéditos*. Brasília: Instituto de Estudos Socio-Economicos.

Viola, Eduardo. 1987. "O Movimento Ecológico no Brasil (1974–1986): Do Ambientalismo à Ecopolítica." *Ecologia e Política no Brasil*, ed. José Augusto Pádua. Rio de Janeiro: Espaço e Tempo / IUPERJ.

———. 1988. "The Ecologist Movement in Brazil (1974–1986): From Environmentalism to Ecopolitics." *International Journal of Urban and Regional Research* 12, no. 2, 211–28.

———. 1992. "O Movimento Ambientalista no Brasil, 1971–1991: Da Denúncia e Conscientização para a Institucionalização e o Desenvolvimento Sustentável." *Ecologia, Ciência e Política*, ed. M. Goldemberg. Rio de Janeiro: Revan.

———. 1996. "A Multidimensionalidade de Globalização, As Novas Forças Sociais Transnacionais e seu Impacto na Politica Ambiental do Brasil, 1989–1995." *Incertezas de Sustentabilidade na Globalização*, ed. L. d. C. Ferreira and E. Viola. Campinas, Brazil: UNICAMP.

Viola, Eduardo J., and Hector R. Leis. 1995. "O Ambientalismo Multissetoral no Brasil para Além da Rio-92: A Desafio de uma Estratégia Globalista Viável." *Meio Ambiente, Desenvolvimento e Cidadania: Desafios para as Ciencias Sociais*, ed. E. J. Viola et al. São Paulo: Cortez.

Vogel, David. 1986. *National Styles of Business Regulation: A Case Study of Environmental Policy*. Ithaca: Cornell University Press.

Waldman, Maurício. 1990. "'Combativo' ou 'De Resultados.'" *Teoria e Debate* 9, January–March, 13–17.

Weinstein, Barbara. 1983. *The Amazon Rubber Boom, 1850–1920*. Stanford: Stanford University Press.

Weyland, Kurt. 1993. "The Rise and Fall of President Collor and Its Impact on Brazilian Democracy." *Journal of Interamerican Studies and World Affairs* 35, no. 1, 1–37.

Wood, Charles, and Roberto Porro, eds. 2002. *Deforestation and Land Use in the Amazon*. Gainesville: University Press of Florida.

Young, Carlos E. F., and Carlos A. Roncisvalle. 2002. *Expenditures, Investment and Financing for Sustainable Development in Brazil*. Serie Environment and Development, no. 58. Santiago: CEPAL / ECLAC and UNDP.

Zhouri, Andréa. 2001. "Árvores e gente no ativismo transnacional: As dimensões social e ambiental na perspectiva dos campaigners britânicos pela Floresta Amazônica." *Revista de Antropologia* 44, no. 1, 9–52.

Zirker, Daniel, and Marvin Henberg. 1994. "Amazônia: Democracy, Ecology, and Brazilian Military Prerogatives in the 1990s." *Armed Forces and Society* 20, no. 2, 259–81.

Zulauf, Werner. 1994. *Brasil Ambiental: Síndromes e Potencialidades*. São Paulo: Konrad-Adenauer-Stiftung.

Newspapers and Newsmagazines

A Tribuna
Diário Grande ABC
Diário Oficial
Estado de São Paulo
Exame
Folha da Tarde
Folha de São Paulo
Gazeta Mercantil
Globo
Isto É (Brazilian newsmagazine)
Jornal da Tarde
Jornal do Brasil
New York Times
Noticias de Poá
Veja (Brazilian newsmagazine)
Washington Post

Interviews Cited

(Positions current as of time of interview. Unless otherwise noted, interviews were done by the authors; names of research assistants who conducted interviews given in parentheses.)

Ab'Saber, Aziz. Professor of geography, University of São Paulo (29 April and 1 May 1991).

Arnt, Ricardo. Journalist (Rio de Janeiro, 1990).

Aveline, Carlos. President, União Protetora da Natureza (UPAN) (São Leopoldo (RS), 3 June 1990).

Barros, Luis Carlos de. Official, Movimento Arte e Pensamento Ecológico (MAPE) (São Paulo, 15 October 1990).

Barros, Ruy de Góes Leite de. Official, Greenpeace (São Paulo, 8 April 1991).

Born, Rubens, and Antonio Carlos Oliveira. Environmental activists, OIKOS (18 October 1990).
Bramble, Barbara. Director of international programs, National Wildlife Federation (Washington, 1990).
Brito Neto, Armand Antonio. Executive director, Pro-Rio, and president, Movimento Pró-Floresta da Tijuca (Rio de Janeiro, 1 August 1990, Biorn Maybury-Lewis).
Capobianco, João Paulo. Director, Fundação SOS Mata Atlântica (São Paulo, 7 August 1990); director, Instituto Sócio-Ambiental (1993).
Cunha, Renato. President, GAMBÁ (Bahia, 5 September 1991).
Feldmann, Fábio. Federal deputy, PSDB-São Paulo (Brasília, 14 September 1991); former state secretary of the environment, São Paulo (São Paulo, 10 July 1999).
Fonseca, Gelson, Jr. Diplomatic aide to President Fernando Collor (Brazília, 1 August 1990).
Gregol, Giovani. Municipal councilor (PT) (Porto Alegre, 6 June 1990).
Gross, Tony. CEDI (Washington, 20 September 1990).
Gualda, Regina. Former coordinator of social communication and environmental education, Secretaria Especial do Meio Ambiente (SEMA) (Brasília, 16 September 1991).
Guttemberg, Anna. União Internacional Protetora dos Animais (São Paulo, between 1985 and 1987, Maria Helena Antuniassi and associates).
Hathaway, David. Alternative Technologies Project (AS-PTA) (Rio de Janeiro, July 1990).
Heckmeir, Luiz Martins. Engineer, Air Quality Division, Fundação Estadual de Engenharia do Meio Ambiente, Rio de Janeiro (Rio de Janeiro, 22 July 1999).
Iglesias, Chico. President, Associação Potigar Amigos da Natureza, and Conama representative (Natal, 2 September 1991).
Lanuza, Cacilda. President, Grupo Seiva (São Paulo, 26 September 1990).
Lisboa, Marijane. Former Secretary of Environmental Quality, Ministry of Environment (MMA), and former Greenpeace official (São Paulo, 12 August 2005).
Lovejoy, Thomas. Official, Smithsonian Museum (March 1992).
Magnannini, Alceu. President, Fundação Brasileira Para a Conservação da Natureza (Rio de Janeiro, 10 August 1990, Biorn Maybury Lewis).
Malerba, Julianna. Projeto Brasil Sustentável e Democrático, FASE, and administrator of Environmental Justice Network (Rio de Janeiro, 5 August 2005).
Marques, Celso. President, Associação Gaúcha de Proteção ao Ambiente Natural (AGAPAN) (Porto Alegre, 7 June 1990).
Marques, Randau. Member, OIKOS, and journalist (São Paulo, 3 April and 29 September 1991).
Milaré, Edis. Procurador de Justiça and coordenator of Curadorias Speciais de Proteção ao Meio Ambiente do Estado de São Paulo (São Paulo, 10 April 1991).
Miller, Carlos. Director, Fundação Vitória Amazônia (Manaus, 29 August 1991).

Muser Filho, [Coronel] Manoel. Companha Nacional para a Defesa e pelo Desenvolvimento da Amazônia (CNDDA) (Rio de Janeiro, 17 July 1991).

Neuhaus, Esther, and Patrícia Bonilha. Executive manager and communications director, Fórum Brasileiro de ONGs e Movimentos Sociais para o Meio Ambiente e o Desenvolvimento (FBOMS) (Brasília, 9 August 2005).

Nogueira Neto, Paulo. Founder of ADEMA (São Paulo, 3 December 1990, Cristina Saliba); first national secretary of the environment in Brazil (11 April 1991).

Oliveira, Antonio Carlos de. Former leader of student anti-nuclear movement (São Paulo, 20 November 1992).

Paiolli, Waldemar. Director, Associação Mundial de Ecologia (13 December 1990, Cristina Saliba); President, Associação Brasileira de Proteção à Natureza (São Paulo, between 1985 and 1987, Maria Helena Antuniassi and associates).

Petrillo, Celso. President, Ação Ecológica, and Conama representative (São Paulo, 22 October 1990).

Rocco, Rogério. Environmental lawyer (Rio de Janeiro, May 2004).

Rodrigues, Délcio. Campaign director, Greenpeace (São Paulo, 15 June 2000).

Sarmento, Jair. Director and executive secretary, Ministry of the Environment (Brasília, 4 August 1999).

Saussuna, Karen. International coordinator, Associação de Combate aos POPs (ACPO) (São Paulo, 3 August 2005).

Schäfer, Wigold. President, Associação de Preservação do Meio Ambiente do Alto Vale do Itajaí, and Conama representative (Santa Catarina, 1990).

Schmidt, Daniel Egon. Manager, Vehicles Inspection and Programs Division, Companhia de Tecnologia de Saneamento Ambiental, São Paulo (São Paulo, 15 July 1999).

Sirkis, Alfredo. Leader of Green Party and Municipal Council member (Rio de Janeiro, 24 November 1992).

Sorrentino, Marcos. President, Sodemap, and official, Apedema (Campinas, São Paulo, 28 August 1990).

Steinbaum, Volf. Sociologist, Coordination Group of Environmental Chambers, Companhia de Tecnologia de Saneamento Ambiental, São Paulo (São Paulo, 15 July 1999).

Svirsky, Enrique. Coordinator, PROAONG, Secretaria de Estado do Meio Ambiente, São Paulo (São Paulo, 7 July 1999).

Valverde, Orlando. President, Campanha Nacional de Defesa e Pelo Desenvolvimento da Amazônia (17 August 1990, Biorn Maybury-Lewis).

Vieira, Liszt. President, Direito e Ambiente (Rio de Janeiro, 4 October 1991); IED (24 November 1992).

Zulauf, Werner. Former president of Companhia de Tecnologia de Saneamento Ambiental, São Paulo (São Paulo, 22 April 1991).

Index

Abellá, Miguel, 188
ABONG (Brazilian Association of NGOs), 100–101, 108
Ab'Saber, Aziz, 77–79, 157–58
Ação Ecológica (Ecological Action), 84
Acre (state), 1–2, 160, 166; distinctiveness of, 183–84. *See also* Rubber tappers
Activist life cycle, 131–35
Adams, Tani, 136–37, 244 n18
Adário, Paulo, 176–77
Afonso, Carlos, 100, 116
Afro-Brazilians, 90
AGAPAN (Association for the Protection of the Natural Environment), 71, 73–74, 82, 85, 89, 91, 105–6, 109, 240 n16
Agenda 21, 240, n12
Agriculture, Ministry of, 29
Air pollution, 2–3; control of, 189, 201–2, 211; industrial, 187; licensing of, 204; neighborhood perceptions of, 212; rationales for combating, 207–8; in São Paulo, 205; sources of, 206; vehicle inspections and, 210. *See also* Cubatão: air pollution in; Rodízio policy
Alencar, Chico, 132
Allegretti, Mary: brokerage role of, 164–65, 171; environmentalism in the Amazon and, 154–55; extractive reserves promoted by, 168–70;

Lutzenberger and, 167; resignation from government of, 170, 181; rubber tappers' movement and, 162; as secretary for Amazônia, 168–70, 178, 241; trajectory of, 168–70, 228
Amapá (state), 2, 144
Amazonas (state), 29, 154, 158
Amazônia: American environmentalists and, 111, 155, 164; blocking and enabling networks in, 230; cattle ranching and, 145, 148–49; colonization of, 145–46; deforestation of, 2, 6, 115; development models for, 146; diversity of, 140, 142, 147, 156; foreign investment in, 149, 151; geopolitics and, 143–47; growing foreign attention to, 162–63; improved data on, 168; international funds for conservation of, 108, 129, 166, 171–72; land tenure and, 146–47; law enforcement in, 151–54; logging in, 149–51; military dimension of, 157–88, 163; nationalism and, 69–70, 143, 177; new environmental organizations and, 95, 172; networks of scientists and, 154; NGOs and, 108; rubber production in, 143; small farmers expelled in, 145, 152–53; soy cultivation in, 175; state and local governments and, 147, 152, 154, 166; violence and criminality in, 1, 2, 151, 153, 177. *See also* Environmen-

Amazônia (*continued*)
  tal campaigns; Greenpeace Amazon campaign
Amazon Studies Institute (IEA), 60
Amazon Working Group (Grupo de Trabalho Amazônico; GTA), 171–72
Amnesty law, 89
Angra I nuclear plant, 85
Anthropologists, 155, 157, 170
Antinuclear movement, 85
Antuniassi, Maria Helena, 64
APEDEMAS: Constituent Assembly and, 47; formation of, 73, 84; 1986 elections and, 93; revitalization of, 132; UNCED and, 118–20
APPN (São Paulo Association for the Protection of Nature), 55, 76–78, 83
Aracruz Celulose, 220
Aramar nuclear research facility, 83
ARENA, 86
Argentina, 4, 83
Arns, Dom Paulo Evaristo, 80
Arnt, Ricardo, 69
Art and Ecological Thought Movement (MAPE), 82, 188
Association against Persistent Organic Pollutants (ACPO), 218–20, 247 n14
Association of Agronomists, 77
Association of Those Exposed to Asbestos (ABRAE), 218
Association of Tropical Biology, 160
Association of Victims of Pollution and Bad Living Conditions of Cubatão, 84, 193–202
Atlantic Forest, 14; exhaustion of, 150, 163
Atlantic Forest Network, 128, 134
Automobiles, 205, 211; emissions from, 209–10

Bahia (state), 28
Belém, 30, 154, 156
Belém-Brasília Highway, 145
Bethlehem Steel, 144
Bhopal disaster, 197
Bierrenbach, Flávio, 77
Biofuels, 207–8
Blocking and enabling strategies, 202, 230
BNDES (National Economic Development Bank), 191
Born, Rubens, 84, 118, 121, 123
Brazil: Amazon and, 113; Argentina and, 83; big projects in, 31; budgetary constraints in, 40; Chamber of Deputies of, 26, 148, 157; debt crisis and, 82; export agriculture and, 8; foreign relations of, 27; German nuclear accord with, 80, 82, 85; hyperinflation in, 95; National Congress of, 26, 87, 108–9, 147; politics and government in, 10, 11, 17, 25–26, 148, 246 n13; UNCED bid by, 113. *See also* Constituent Assembly
Brazil-Germany Nuclear Accord, 80, 82, 85
Brazilian Architects' Institute, 78
Brazilian Association of Sanitary Engineers, 29, 77
Brazilian Bar Association, 84
Brazilian Botanical Society, 77–78
Brazilian Foundation for the Conservation of Nature (FBCN), 66–68
Brazilian Institute for Forestry Development, 29, 68, 78–79, 104, 239 n2; IBAMA and, 36
Brazilian Institute of the Environment and Natural Resources. *See* IBAMA
Brazilian Network on Multilateral Development Institutions, 170–71
Brazilian NGO Forum, 115, 118, 120–25, 128–29, 175
Brazilian Society of Landscape Architects, 78
Brazilian Society of Physicists, 80–81

Buarque, Cristovam, 6
Budgets, nondisbursement of, 175
Cabral, Bernardo, 48
Calha Norte, 163
California, 203–4
Capobianco, João Paulo, 228; changing environmental movement and, 130; environmentalist strategy and, 43; Ministry of Environment and, 104–5, 243 n3; pro-Juréia movement and, 81, 102; SOS-Mata Atlântica and, 102–4
Carajás, 144–45, 244 n1
Cardoso, Fernando Henrique, 39, 41, 79, 168, 169
Catholic Church: Amazon and, 155, 160, 161, 179, 243 n1; as associational space, 154; Cubatão and, 196; environmental campaigns and, 77, 86; NGO sector and, 99–100
Caucaia airport, 72, 75–79
CDPC (Commission to Defend the Community Patrimony), 77–78, 83–84, 157–58
CEDI (Ecumenical Center for Documentation and Information), 118, 121
Center for Our Common Future, 117–18, 123
Center for Support to Traditional Peoples, 167
Centrão, 49–50
CETESB (Environmental Sanitation Technology Company): air pollution standards and, 209; Cubatão and, 190, 197–98, 200; community participation and education programs of, 197; creation of, 188; democracy-technocracy debates within, 189; manipulation of, 225; Rodízio and, 213, 216; vehicle inspections and, 210; vulnerability to political workplace contaminants and, 219

CIEC (Interstate Ecological Coordination for the Constituent Assembly), 88, 93–95
Classifications, of environmentalists, 64–66
Climate change, 186, 203, 206
CNDDA (National Campaign for the Defense and Development of the Amazon), 69–70, 145, 245
Codim, 145. *See also* Grande Carajás program
Collor de Mello, Fernando, 43, 59, 105, 114, 127, 227
Colonization and Land Reform Institute (INCRA), 29, 146, 148
Commission to Defend the Community Heritage (CDPC), 77–78, 83–84, 157–58
Communist Party, Brazilian, 70
Companhia Vale do Rio Doce (CVRD), 145. *See also* Grande Carajás program
CONAMA (National Environmental Council): Cardoso government and, 39; Collor government and, 37, 43; creation and expansion of 32–36, 240 n6; environmental impact statements and, 36, 201–2; environmentalists and, 42–45; vehicular pollution and, 209, 210
CONDEPHAAT, 78–79
Conduct Adjustment Agreements, 54–55
Conference of Ecologists and Pacifists (1985), 87
Conferences, 6
CONSEMA (State Environmental Council), 202–4, 240 n12
Conservation International, 104, 173
Conservation organizations: developmentalist nationalism and, 69–70; electoral politics and, 88, 91; environmental ideas and, 7; history of, 7, 66–70

Conservation units: Constituent Assembly and, 50; creation and protection of, 2, 173–74; debate over direct use of, 68, 246 n14; ecological stations as, 29–30; Environmental Protection Areas (APAs) as, 32; Stang's murder and, 181

Constituent Assembly: campaign for environmental chapter of, 46, 48–50; electoral politics and, 84, 86–87, 90–91, 95. *See also* National Front for Ecological Action in the Constituent Assembly

Constitution of 1988, 14, 34, 147, 218, 226–27

CONTAG (National Confederation of Agricultural Workers), 160–61

COSIPA (National Steel Company), 191

Cosmopolitans, 9

Cotia, 76–78

Councils, environmental, 34, 43, 227, 228. *See also* CONAMA; CONSEMA

Covas, Mário, 211, 215

Cruzado Plan, 95

Cubatão: air pollution in, 186, 189, 225; Association of Victims and, 84, 193–202; cleanup in, 1, 197; congenital deformities in, 191–92; COSIPA and, 191, 197, 201; environmental justice and, 217; Inter-ministerial Commission and, 194; major industries in, 246 n4; as national disaster, 195; as national security area, 191, 196; public opinion and, 199; as steel and petrochemical pole, 190–92; as Valley of Death, 1, 192–93; zoning decree and, 194

Cuiabá, 150

Curitiba, 205–6

CUT (Central Workers' Union), 160, 244 n17; environmental justice and, 217

Debt for Nature Swaps, 103, 243 n5

Decision making, 129; demands for participation in, 43, 218, 225, 240 n12; environmental impact assessments and, 189; limits of, 43, 221; power to influence state and, 2, 228–28. *See also* Councils, environmental

Deforestation, 2, 6, 115, 149–50, 169, 180–81

Democratization: decentralization and, 32; environmental protection and, 11, 225; participation and, 96; social movements and, 11–12; politics of, 35, 86–89

Development model, 142, 190, 200

Diffusion, 4–5, 8, 216, 229

Disasters, 197, 222; nuclear, 82

Domestic-international collaboration: on Amazon projects, 166–67; on climate change, 203–4; Fundação Vitória Amazônia and, 172–74; Greenpeace International and, 176; international scientific networks and, 157; monitoring group for BR-163 and, 180; PPG-7 and, 171–72; Rede Brasil and, 170–71; rubber tappers and, 155, 157–65; Stang's murder and, 181

Domestic-international linkages, x, 7, 99; Acre rubber tappers and, 111, 163–67; Brazilians in Washington and, 173; early instances of, 30; environmental justice and, 216; growing collaboration in, 171–72; mechanisms of, 9; pros and cons of, 8; Rede Brasil and, 171; UNCED and, 123–27

Drug trafficking, 151–52

Earth Summit. *See* UNCED

Ecological risk, 190

Eco-93. *See* UNCED

Eco-pacifism, 95

Elections: environmentalists and, 87–92, 226; of 1982, 196
Eletrobrás, 85
Environmental campaigns, 73; antinuclear, 79–83; Caucaia airport and, 72, 75–79; development of networks for, 74–75; timber concessions and, 72, 158–60. *See also* National Front for Ecological Action in the Constituent Assembly
Environmental Defense Fund, 171
Environmental education, 31, 128, 200; Rodízio Plan and, 213
Environmental impact, 199, 202
Environmental Impact Statements, 36
Environmental Justice, 1, 109, 186, 216–18, 229
Environmental Justice Network, 1, 3, 200, 216, 218–21
Environmental legislation, 25, 31–33, 36, 38–39
Environmental licensing, 202–3
Environmental movement: apparent stalemate of, 129; conservationism vs., 64; distinct visions of, 82; electoral politics debated within, 87–92, 226; identity of, 101, 132; insider-outsider relations in, x; leftist politics in, 100, 124; Lula policies protested by, 180–81; "new environmentalism" and, 70–74; North-South divisions and, 117; political nature of, 10, 110; problem of leadership renewal in, 130; protest vs. results orientation in, 101–2; Rodízio Plan and, 213, 215; in São Paulo, 75–84; science and, 67–68, 72–73; self-evaluation of, 130; socio-environmentalism and, 109–11; tensions in, 128. *See also* National Front for Ecological Action in the Constituent Assembly
Environmental organizations: environmental education and, 128; funding for, 73, 98, 103–5, 107–8; legal strategies of, 55–56; in multi-organizational field, 12; politicians mistrusted by, 87; professionalization of, 83–85, 99–109, 130; services provided by, 130; size of, 106–7; state agencies' collaboration with, 96, 128; UNCED and, 47, 49, 126–27; urban focus of, 186
Environmental policy: under Cardoso government, 39–40; citizens' participation in, 42; under Collor government, 37–38; enforcement problems and, 25; federal financing of, 40–41; frequent revisions of, 38–40; international financing of, 41–42; judicialization of, 45; under Lula government, 180–81; military regime and, 24; natural resources and, 39; regional disparities in resources for, 15; under Sarney government, 36–37, 87–89. *See also* CONAMA; IBAMA; SEMA
Environmental politics: in broader context, x, 229; complexity of, 3; democratic transition and, 10–13; federalism and, 10, 13–16
Environmental professions, 130, 134
Environmental Sanitation Technology Company. *See* CETESB
Environmental Secretariat. *See* SEMA
Ethanol, 207–8
European Union, 171
Exceptionalism, Brazilian and American, 10
Exiles, return of, 64, 87, 89, 93, 99
Extractive reserves: Allegretti's promotion of, 168–70; development of idea of, 156, 162; economic viability of, 168; as enabling strategy, 162; legal instrument for, 165, 167. *See also* Rubber tappers

FASE (Federation of Organs for Social and Educational Assistance), 100, 108, 121, 218
FBCN (Brazilian Foundation for the Conservation of Nature), 66–68
FBOMS (Forum of Brazilian NGOs for Environment and Development), 128–29. *See also* Brazilian NGO Forum
Federalism, 10, 13–16; under 1988 Constitution, 14, 147; opportunities for political action and, 92–93
Feldmann, Fábio: Constituent Assembly and, 46–48, 91, 226; Cubatão and, 55, 194; environmental organizations mobilized by, 213; on networks, 221; networks and political project of, 59, 95; OIKOS and, 84, 102; political and social focus of, 211, 228; Rodízio plan and, 212–15, 247 n12; as state secretary of the environment, 211–15; workplace contamination and, 219
Feminism, 93, 95
FINEP (Funding Agency for Studies and Projects), 29, 239 n3
Flexible-fuel cars, 208
Ford Foundation, 101, 104, 221
Forest Code, 66, 78, 149, 169
Forest Concession Law, 245 n3
Forest management, 157–58, 160, 163
Forum of Brazilian NGOs for Environment and Development, 128–29. *See also* Brazilian NGO Forum
Franco, Itamar, 39
Friends of the Earth, 129, 180
Frontier, 153–54, 163
Fundação SOS–Mata Atlântica, 102–5, 117, 128
Fundação Vitória Amazônica (FVA), 172–74

G-7 Pilot Project for Conservation of the Brazilian Rainforest (PPG-7), 2, 108, 129, 166, 171–72
Gabeira, Fernando, 89, 91, 93–95, 132
Geisel, Ernesto, 71, 79, 146, 158
Germany, 64, 80, 82, 85
Global Forum, 118, 120, 124–25, 127–29, 217
Goiás (state), 146
Goldman, Alberto, 79
Goree, Langston James, III (Kimo), 116
Grande Carajás program, 144–45, 244 n1
Green Collective, 89
Green Front, 47–48, 50, 168
Green List, 88, 93–95
Green Parties, 90, 92
Green Party, Brazilian, 12, 48, 168, 226; antinuclear activism and, 85; electoral record of, 111; founders of, 89; homosexual rights and, 92; lake event and, 95; libertarianism and, 90; origins of, 88–93; self-governing ecological socialism and, 91, 242 n12; student wing of, 132
Greenpeace, 99, 129, 229, 244 n20; air pollution campaign of, 201, 213, 215; Amazon campaign of, 137–38, 175–77; distinctive culture of, 135; international organization of, 135
Gregol, Giovanni, 105
Gross, Anthony: brokerage role of, 164, 171; NGO Forum and, 116, 121; rubber tappers and, 117–18; 155, 161; trajectory of, 117
Group for the Study of Air Pollution, 201
Grüde, 132–33
Grupo Seiva de Ecologia, 83
GTA (Amazon Working Group), 171–72
Guerreiro, Saraiva, 85
Guimarães, Ulysses, 48

Hecht, Susanna, 162
Heinrich Böll Foundation, 221
Highways: Belém-Brasília and, 30, 143; under Cardoso and Lula governments, 180; Grande Carajás program and, 144; monitoring group for BR-163 and, 180, 227; resistance toward, 169; Rondônia and, 165; World Bank and, 164
Hiroshima Day, 81–82
Hudson Institute, 69, 145
Human rights, 1, 5, 218

IBAMA (Brazilian Institute of the Environment and Natural Resources), 114, 135, 148, 150, 240 n8; Center for Support to Traditional Peoples and, 167; creation of, 36; under SEMA, 37
IBASE (Brazilian Institute of Social and Economic Analyses), 99, 108, 120–21, 173
IBDF, 29, 36, 68, 78–79, 104, 239 n2
INCRA (Colonization and Land Reform Institute), 29, 146, 148
Indigenous peoples, 68, 90, 108
Inequality, 12, 51, 224; in access to transportation, 205; environmental injustice as form of, 217–19; ethanol production and, 208; racial, 206
Informal politics, 16–21
INPA (Institute for Research on the Amazon), 156, 173
Institute for Energy and Nuclear Research (IPEN), 83
Institutions, 16, 224–27
Integrated System for Monitoring, Licensing and Control of Deforestation, 169
Inter-American Development Bank, 164, 168
International Facilitating Committee, 117, 122–24. See also Brazilian NGO Forum

International foundations, 104, 129
International Institute for Hiléia Amazônica, 157
International Labor Organization (ILO), 220
International Monetary Fund (IMF), 149
International Union for the Conservation of Nature (IUCN), 27, 30, 58
International Working Group, 114
Interstate Ecological Coordination for the Constituent Assembly (CIEC), 88, 93–95
ISA (Social-environmental Institute), 104

John D. and Catherine T. MacArthur Foundation, 104, 170
Juréia Ecological Station, 80

Klabín, Roberto, 84, 102
Kubitschek, Juscelino, 143

Land tenure, 146
Lanuza, Cacilda, 83
Law: deficiencies of judiciary and, 52; enforcement of, 1–3, 18, 25, 51, 184–85, 214; judicialization of environmental conflicts and, 45; police and, 148; public interest lawsuits and, 48, 55; rubber tappers and, 161; state building and, 226. See also Ministério Publico
Law to Protect Water Sources, 78
Lazzarini, Walter, 77, 84
Legal Amazônia, 140, 148
Legislative Assembly, 213
Legislative process, 17–18
Leroy, Jean-Pierre, 218
Leverage, 9
Lovejoy, Thomas, 30, 58
Lula da Silva, Luiz Inácio: conservation and, 169, 227; environmentalists and,

Lula da Silva, Luiz Inácio (*continued*)
134, 178, 180–81, 219; environmental spending and, 40; in 1989 election of, 93, 105; road projects and, 169; rubber tappers and, 161, 178
Lustoso, Caio, 91
Lutzenberger, José, 76; AGAPAN and, 73–74, 89; Amazon and, 37, 60; 158, 167; antinuclear movement and, 80; Collor and, 114; dismissal from government of, 115; ecological manifesto published by, 74; as environmental secretary, 37, 60; 105; life of, 240 n9

Magnannini, Alceu, 67
Maluf, Paulo, 81, 203, 210, 247 n10
Manaus, 156, 160
Marinho, Roberto, 28
Marinho, Rogério, 28
Marques, Celso, 74
Marques, Randau, 76, 84, 194
Martins, Paulo Egydio, 76, 78–79
Marx, Burle, 78
Mato Grosso (state), 150
MDB (Brazilian Democratic Movement), 79, 86, 89, 157–58, 192
Media, 28, 76, 78, 84, 102; Cubatão and, 192, 199; environmentalists' use of, 108; Mendes and, 111; Rodízio plan and, 212–14; UNCED and, 126
Médici, Emílio Garrastazú, 71, 152, 157
Mendes, Francisco (Chico), 1, 2, 109, 155, 227; Allegretti and, 162, 168; Marina Silva and, 178–79; murder of, 110–11, 155, 165, 172; in United States, 164
Menezes, Nanuza, 77–78
Mesquita, Antonio Carlos, 78
Mesquita, Júlio, 28
Mesquita, Rodrigo, 84, 102
Milaré, Edis, 54
Miller, Carlos, 173

Minas Gerais (state), 145–46
Minc, Carlos, 89, 93, 94, 110, 132
Ministério Público, 34, 45, 52–57, 150, 239 n11; Conduct Adjustment Agreements and, 54–55; Fiat case and, 209; Greenpeace meetings with, 215; near-absence in Pará of, 148; workplace contamination lawsuits and, 219–20
Ministry of Aeronautics, 77
Ministry of the Environment, 2; budget of, 40–42; institutional configurations of, 38–40, 149, 241, 227
Montoro, Franco, 79, 196–98, 203
Morelli, Mauro, 77
Movement against Nuclear Reactors, 81
Movement in Defense of the Amazon, 158–60
Movement of Those Affected by Dams (MAB), 218. *See also* Environmental Justice Network
Movimento de Arrecadação Feminina, 78, 81
Multilateral Bank Campaign, 111, 164, 171
Multilevel governance, 3, 15–16, 230
Museu Paraense Emilio Goeldi, 156

National Agrarian Reform Policy, 168
National Alcohol Program (Proalcohol), 207–8
National Confederation of Industry, 33
National Conference of Physics Students, 80
National Council of Rubber Tappers (CNS), 146
National Environmental Fund, 42, 135
National Front for Ecological Action in the Constituent Assembly (National Front, Green Front), 47–48, 50, 168
Nationalism, 70, 108; Amazônia and, 69–70, 112, 143, 156–57, 159; conservation and, 96; foreign environmental NGOs and, 177; of the left, 69–70,

145, 245 n2. *See also* Sovereignty and environment
National Parks, World Conference on, 68
National Sanitation Policy, 26
National Security Law, 161
Nefussi, Nelson, 188
Networks, 16, 116; Amazônia and, 182; blocking and enabling strategies of, 11, 19–20, 75, 81, 83, 85–86, 101–2, 162, 170, 188, 199, 202, 203, 221, 228; costs of, 241 n2; creation of, 20; FVA and, 173; as organizational form, 3; Rede Brasil and, 171; social, 29, 99–100, 228; state capacity and, 18, 19, 29; transnational advocacy and, 6
Neves, Tancredo, 35
New environmentalism, 70–75, 96
NIMBY (Not in My Backyard), 219
Nogueira Neto, Paulo: airport protest and, 76, 78; conservation movement and, 69; Constituent Assembly and, 48; extractive reserves backed by, 165; international contacts of, 30; personal networks used by, 28–29, 58; resignation of, 35; as secretary of environment, 24, 27–29
Nongovernmental organizations (NGOs), 98–101, 106–8
Norm diffusion, 4–5, 10
Nossa Natureza, 128, 149

OIKOS, 46, 55, 84–85, 117, 194
Oliveira, Antonio Carlos de (Tonhão), 80–82
Os Verdes, 133–34
Oxfam, 117, 161

Padua, José Augusto, 66, 176.
Pádua, Maria Teresa Jorge, 48, 58–59, 68, 102, 239 n2
Paiolli, Valdemar, 76, 78
Pantanal, 14, 163

Pará (state), 1, 106, 148, 151, 169, 227
Paraná, 146, 156, 162
Path dependence, 223
Pavan, Clodowaldo, 194
Payoffs, in access to funding, 9, 229
PDS (Democratic Social Party), 86
Permanent Campaign against the Transfer of Toxic Wastes between States of the Federation, 1, 3, 200, 216, 218–21
Persuasion, 9
Peru, 152
Petrobrás pipeline explosion, 198
Pinheiros, Wilson, 161
Pinto, Lúcio Flávio, 151–52, 178, 245 n3
Plan for Prevention of Extreme Episodes of Air Pollution, 200
PMDB (Party of the Brazilian Democratic Movement), 86, 89, 93–94, 196, 226
Police, 148, 150
Policy process, 17, 18
Political parties, 86, 226
Pollution: as public concern, 187; as social problem, 193. *See also* Air pollution
Pollution control, 187, 189
Post-materialism, 93
Poverty, 110
PPG-7 (G-7 Pilot Project for Conservation of the Brazilian Rainforest), 2, 108, 129, 166, 171–72
PROCONVE (Program for Control of Air Pollution from Automobiles), 209
Professionalization of environmental organizations, 83–85, 97; UNCED and, 120–21, 128
Pró-Jureia Movement, 81
Pró-Rio Coalition, 117, 118, 244 n15
PSB (Brazilian Socialist Party), 93
PSDB (Brazilian Social Democratic Party), 226

Index 281

PT (Workers' Party), 72, 86, 89–93, 95, 105, 110–11, 155, 161, 179, 196, 213, 226
Public interest lawsuits, 48, 55
Public opinion, 28–29, 199
PV. *See* Green Party

Quadros, Janio, 27

Radio Eldorado, 104
Rede Brasil (Brazil Network on Multilateral Financial Institutions), 170–71
Reis, Arthur Cesar Ferreira, 70
Rhodia, 1, 192, 219–20
Rio Grande do Sul (state), 73, 92–93, 105, 109, 168
Risk Contracts Campaign, 32, 72, 150–51, 158–60, 163–64
Rocco, Rogério, 119, 131–35, 228
Rodízio policy, 211–16, 247 n12
Rondon, Cândido Mariano da Silva, 69, 143
Rondônia (state), 106, 144, 146, 148, 150, 152, 161, 164, 166, 167
Rotary Club, 64, 75-77, 83, 241 n4
Rubber tappers, 257 n7; alliances of, 111, 155, 160–61; environmental frame adopted by, 165; international funding for, 166; National Council of, 146; organization and politicization of, 108, 183; Rubber Tapper Project (Projeto Seringueiro), 168

Salvador, 28
Santarém, 144
Santos, 28
Santos, Dojival Vieira dos, 198
São Paulo: as environmental policy innovator, 2, 188; land grabbers from, 146; legislative underrepresentation of, 148; quality of life in, 187
São Paulo Association for the Protection of Nature (APPN), 55, 76–78, 83
São Paulo Engineering Institute, 78

Saragoussi, Muriel, 172–74, 181
Sarney Filho, José, 42, 50, 95, 112, 149, 168
SBPC (Brazilian Society for the Promotion of Science), 75, 81; Amazônia and, 68, 247 n5; Constituent Assembly and, 46; in Cubatão struggle, 194–96
Scale, 9
Schinke, Gert, 105
Schwartzman, Stephan, 69, 170
Science: alternative paradigms in, 72–73; authority of, 195, 247–48 n5; idea of nation and, 245 n5; politics and, 85
Scientists, 154–57
Secretariat of Health (São Paulo), 187–88
Secretariat of the Environment (SEMA): creation of, 27, 229; extractive reserves and, 114–15, 227; institutional location of, 35–37; Lutzenberger and, 37, 60; Nogueira Neto and, 24, 27–29; weakness during military regime of, 27, 30–33, 35
Silva, Marina, 156, 170, 179–81, 225
Sirkis, Alfredo, 89, 94, 132, 168, 242 n11
SISNAMA (National System for the Environment), 32–34, 128, 240 n6, 240 n8
SIVAM (System for Surveillance of the Amazon), 163
Smithsonian Institution, 30
Social movements, 12
Socio-environmentalism, 10, 72, 98, 108; Amazônia and, 174, 178; appearance of, 109; environmental justice and, 216; equity and participation as elements of, 13; foreign conservationist skepticism toward, 176; nonresonance in urban settings of, 109, 229; tensions within, 121–22
Sorocaba, 83
SOS (Fundação SOS-Mata Atlântica), 102–5, 117, 128

Souza, Herbert "Betinho" de, 100
Sovereignty and environment, 6, 113, 122, 159, 229. *See also* Nationalism
Soybean cultivation, 149, 175, 180
SPVEA (Agency for the Valorization of the Amazon), 143
Stang, Dorothy, 1, 177; murder of, 181, 217, 227
State building, 141
State employment, 228
State Program for Climate Change, 204
State Secretariat of the Environment, 200, 209, 211–15
State-society networks, 162, 213, 228, 230
Stockholm Conference on the Human Environment (1972), 4, 6, 112–13, 187, 229
Stockholm Convention on Persistent Organic Pollutants, 218–20, 247 n14
Subcontracting, 148
SUDAM (Agency for Amazonian Development), 144; forest policy and, 150, 158
SUSAM (Environmental Sanitation Agency), 188

Technical roles, 102, 197, 203, 222
Technocracy vs. democracy debates, 189
Tietê river campaign, 104
Timber concessions, 32, 72, 150–51, 158–60, 163–64
Timber extraction, 2, 149–51
Trajectories of individuals, 7, 21; of Allegretti, 168–70, 228; Green Party and, 94; of Gross, 117; political skills and, 57–60; of Silva, 179
Transnational relations, 229. *See also* Domestic-international linkages
Transportation, 204–5, 214
Treaty of Tlatelolco, 83

UNCED (United Nations Conference on Environment and Development), 99, 102, 104, 227, 229; Brazilian environmental organizations affected by, 127–28; Brazilian stance in, 112–15; Brazil's bid to host, 113–14; as diplomatic opportunity, 114–15; logistical preparations for, 109; NGO participation in, 118; preparatory conferences for, 121. *See also* Global Forum
União Ecológica, 83
United Nations Conference on the Human Environment (1972), 4, 6, 112–13, 187, 229
United Nations Development Program, 188
U.S. Congress, 163–64
U.S. Steel, 145
University of São Paulo, 77
Urban environmentalism, 217–18
Urban problems, 186–87

Vargas, Getúlio, 51, 143, 144, 190
Vehicular emissions, 209–10
Vianna, Aurelio, 170
Vianna, Jorge, 156, 178, 179
Vieira, Liszt, 89, 94
Vila Parisi, 193, 200
Vila Socó, 193, 198
Viola, Eduardo, 65
Violence, 1, 151–53, 181; privatization of, 217
Volkswagen, 146
Voluntarism, 106, 120
Von der Weid, Jean Marc, 100

World Bank, 36, 164, 167, 171, 188; Mendes and, 165
World Health Organization, 188, 201
World Wildlife Fund, 30, 99, 104, 129, 149, 173

Xapurí, 168

Yanomami, 114

Zulauf, Werner, 197, 247 n10

Kathryn Hochstetler is a professor of political science at the University of New Mexico.

Margaret E. Keck is a professor of political science at Johns Hopkins University.

Library of Congress Cataloging-in-Publication Data

Hochstetler, Kathryn, 1962–
Greening brazil : environmental activism in state and society /
Kathryn Hochstetler and Margaret E. Keck.
p. cm.
Includes bibliographical references and index.
ISBN 978-0-8223-4048-5 (cloth : alk. paper) —
ISBN 978-0-8223-4031-7 (pbk. : alk. paper)
1. Environmentalism—Brazil. 2. Environmentalism—Political aspects—Brazil.
3. Green movement—Brazil. 4. Brazil—Environmental conditions.
I. Keck, Margaret E. II. Title.
GE199.B6H63 2007
333.720981—dc22
2007004080